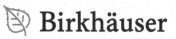

John W. Dawson, Jr.

Why Prove it Again?

Alternative Proofs in Mathematical Practice

with the assistance of Bruce S. Babcock
and with a chapter by Steven H. Weintraub

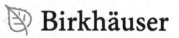 Birkhäuser

John W. Dawson, Jr.
Penn State York
York, PA, USA

ISBN 978-3-319-34967-1   ISBN 978-3-319-17368-9   (eBook)
DOI 10.1007/978-3-319-17368-9

Mathematics Subject Classification (2010): 00A35, 00A30, 01A05, 03A05, 03F99

Springer Cham Heidelberg New York Dordrecht London
© Springer International Publishing Switzerland 2015
Softcover reprint of the hardcover 1st edition 2015

Printed on acid-free paper

Springer International Publishing AG Switzerland is part of Springer Science+Business Media (www.
springer.com)

To Solomon Feferman,
friend and mentor,
who suggested I write this book

Preface

This book is an elaboration of themes that I previously explored in my paper "Why do mathematicians re-prove theorems?" (Dawson 2006). It addresses two basic questions concerning mathematical practice:

1. What rationales are there for presenting new proofs of previously established mathematical results?
   and
2. How do mathematicians judge whether two proofs of a given result are essentially different?

The discovery and presentation of new proofs of results already proven by other means has been a salient feature of mathematical practice since ancient times.[1] Yet historians and philosophers of mathematics have paid surprisingly little attention to that phenomenon, and mathematical logicians have so far made little progress in developing formal criteria for distinguishing different proofs from one another, or for recognizing when proofs are substantially the same.

A number of books and papers have compared alternative proofs of particular theorems (see the references in succeeding chapters), but no extended general study of the roles of alternative proofs in mathematical practice seems hitherto to have been undertaken.

Consideration of particular case studies is, of course, a necessary prerequisite for formulating more general conclusions, and that course will be followed here as well. The aim is *not*, however, to arrive at any *formal* framework for analyzing differences among proofs. It is rather

a) to suggest some *pragmatic criteria* for distinguishing among proofs, and
b) to enumerate reasons *why* new proofs of previously established results have so long played a prominent and esteemed role in mathematical practice.

---

[1] Wilbur Knorr, e.g., noted that "multiple proofs were frequently characteristic of pre-Euclidean studies" (Knorr 1975, p. 9).

Chapter 1 addresses the first of those aims, following clarification of some pertinent logical issues. Chapter 2 then outlines various purposes that alternative proofs may serve. The remaining chapters provide detailed case studies of alternative proofs of particular theorems. The different proofs considered therein both illustrate the motives for giving alternative proofs and serve as benchmarks for evaluating the worth of the pragmatic criteria in terms of which they are analyzed.

York, PA, USA   John W. Dawson, Jr.

References

Dawson, J.: Why do mathematicians re-prove theorems? Philosophia Mathematica (III) **14**, pp. 269–286 (2006)

Knorr, W.: The Evolution of the Euclidean Elements. A Study of the Theory of Incommensurable Magnitudes and its Significance for Early Greek Geometry. Reidel, Dordrecht (1975)

Acknowledgments

I am indebted to Solomon Feferman, Akihiro Kanamori, and two anonymous reviewers for suggesting improvements to earlier versions of chapters 1–8. I thank the participants in the Philadelphia Area Seminar on the History of Mathematics, as well as Andrew Arana and Jeremy Avigad, for their enthusiasm for and encouragement of this endeavor. Above all, I am grateful to Bruce Babcock, for preparing the illustrations throughout this book and for his careful copy-editing, and to Steven H. Weintraub, for several enhancements to chapter 6 and for contributing chapter 11 on proofs of the irreducibility of the cyclotomic polynomials. Finally, I thank Anthony Charles and the rest of the Birkhäuser production staff for their cordial and efficient efforts to transform my manuscript into print.

Contents

Chapter 1
Proofs in Mathematical Practice

Before proceeding to consider the questions posed in the Preface, it is necessary to clarify some logical issues. Paramount among them is the question: **What is a proof?**

The notion of what constitutes a proof in a *formalized* theory is perfectly precise: It is a finite sequence of well-formed formulas, the last of which is the statement to be proved and each of which is either an axiom, a hypothesis, or the result of applying one of a specified list of rules of inference to previous formulas of the sequence. That notion of proof, central to mathematical logic, has led to important advances in the understanding of many foundational and metamathematical issues, and it is widely believed that all current mathematical theories can be formalized within the framework of first-order Zermelo-Fraenkel set theory. Nevertheless, formal proofs are not the focus of the present inquiry.

One reason they are not is that formal proofs have not yet become the stuff of mathematical practice:[1] The notion of a formalized theory is of very recent origin, and strictly formal proofs appear almost exclusively in texts on logic and computer science, not in ordinary mathematical discourse. In addition, mathematicians do not ordinarily resort to formalization in order to judge whether two (informal) proofs that deduce the same conclusion from the same premises are essentially the same. Rather, it is usually easy to tell, on informal grounds, whether such proofs are essentially different or merely variants of one another. Intuitively, they are different if they employ *different concepts or tactics*.

Furthermore, as Yehuda Rav has stressed, mere *expressibility* within the language of set theory does not imply that "all ... current *conceptual proofs* can be formalized as *derivations*" within that theory (Rav 1999, p. 20, fn 20). That is so in part because informal proofs often involve "topic-specific moves" that "have no independent

---

[1]That may well change soon, however, given that computer proofs of such major results as the Four-color Theorem, the Prime Number Theorem, and the Jordan Curve Theorem have now been obtained.

© Springer International Publishing Switzerland 2015
J.W. Dawson, Jr., *Why Prove it Again?*, DOI 10.1007/978-3-319-17368-9_1

logical justification," but serve rather as conceptual "bridges between the initially given data, or between some intermediate steps, and subsequent parts of the argument" (*Ibid.*, p. 26). And, as the logician Jon Barwise noted,

> current formal models of proof are severely impoverished .... For example, ... proofs where one establishes one of several cases and then observes that the others follow by symmetry considerations [constitute] a perfectly valid (and ubiquitous) form of mathematical reasoning, but I know of no system of formal deduction that admits of such a general rule. (Barwise 1989, p. 849)

· That is not to say, however, that formal methods are of no use in comparing informal proofs. Indeed, formal methods may sometimes help to clarify whether two proofs of a statement really establish the same result. For example, one primary motive for presenting alternative proofs is to eliminate *superfluous hypotheses* (as in Goursat's improvement of Cauchy's theorem on integrals of analytic functions) or *controversial assumptions* (such as the Axiom of Choice). But **should a proof based on fewer or weaker hypotheses be regarded as establishing the *same* or a *stronger* theorem?**

Suppose, for example, that a theorem T is first proved from a set of hypotheses H, but that later a proof of T is found that employs as hypotheses only some proper subset P of H. Have we given two different proofs of T, or have we proved in the first case the implication $\bigwedge H \Rightarrow T$ (where $\bigwedge H$ denotes the conjunction of all the hypotheses in H) and in the second case the implication $\bigwedge P \Rightarrow T$ (a stronger result)? Formally, the distinction is merely a matter of perspective, since the Deduction Theorem for first-order logic establishes the equivalence of $A \vdash T$ and $\vdash \bigwedge A \Rightarrow T$ for any set A of non-logical axioms. To avoid confusion, however, the first perspective will be adopted here, according to which it is the *proof*, rather than the *theorem*, that has been strengthened; and for precision, all and only those premises that are actually employed in a proof will be regarded as the hypotheses of the deduction. It then follows that two proofs of the same theorem based on logically *inequivalent* sets of premises must be regarded as different, since the totality of contexts (models of the premises) in which one proof is valid is not the same as those in which the other is.

In some cases, formal considerations may lead to conclusions that differ from those arrived at informally. For example: **If a statement S is known to imply a statement T, should a proof of S ipso facto be regarded as a proof of T?** Informally, the answer ought to be "no" in general, since the proof that S implies T may itself be highly non-trivial.[2] Formally speaking, however, it follows from (the strong form of) Gödel's completeness theorem for first-order logic that if S and T are *any* two theorems provable from some first-order set of axioms A, then so is

---

[2]Nevertheless, as a colleague has rightly noted, if T is a statement whose proof has long been sought and the implication $S \Rightarrow T$ has already been established, one who proves S is often said to have proved T. For example, Andrew Wiles proved the Taniyama conjecture, but is often said to have proved Fermat's Last Theorem.

$S \Rightarrow T$: For both S and T, and therefore also the equivalence $S \iff T$, must hold in every structure that satisfies the axioms, and so, by Gödel's theorem, $S \iff T$ must be provable from the axioms A. Thus, given a proof of S, applying *modus ponens* to the proofs of S and of $S \Rightarrow T$ would yield a proof of T, even though, from a semantic standpoint, S might be utterly irrelevant to T.

Such proofs do not occur in actual mathematical practice. But what if the implication $S \Rightarrow T$ is more easily seen (especially if S is harder to prove than T)? For example, should a proof that the series of reciprocals of the primes diverges, or a proof of Bertrand's Postulate, be deemed a proof of the infinitude of the primes? Should a proof of the Pythagorean Theorem be deemed a proof of the Law of Cosines? The answer is somewhat subjective, and it seems impossible to draw a clear-cut boundary. In Aigner and Ziegler's *Proofs from the Book* (Aigner and Ziegler 2000), for example, a proof that the aforementioned series diverges is included among proofs of the infinitude of the primes, but a proof of Bertrand's Postulate is given in a separate section. As for the Law of Cosines, a direct proof of it — that is, one that does not employ the Pythagorean Theorem[3] — certainly establishes the Pythagorean Theorem as a special case. But in practice, as in Euclid, the Pythagorean Theorem is proved first and then used as a tool to prove the Law of Cosines — a proof which, though relatively straightforward, is not trivial. Thus, although logically equivalent to the Law of Cosines, in practice the Pythagorean Theorem exhibits a certain *conceptual primacy*. It does not seem proper, therefore, to regard a direct proof of the Pythagorean Theorem as a proof *per se* of the Law of Cosines; rather, one may distinguish proofs of the latter according to whether they do or do not rely on the Pythagorean Theorem.

We do not, then, eschew the use of formal *methods* in the analyses to be undertaken here. But the formal model of what constitutes a proof is an abstraction designed to "provide an explanation of … [how] an informal proof is judged to be *correct* [and] what it means for a [mathematical statement] to be a *deductive consequence*" of certain other statements.[4] For those purposes, the formal model of proof serves very well. It seems ill suited, however, for dealing with the broader sort of questions considered here.

Accordingly, the term 'proof' will here be taken to refer to an informal argument, put forward to convince a certain audience that a particular mathematical statement is true — and, ideally, to explain *why* it is true — an argument that is subsequently accepted as valid by consensus of the mathematical community. As such, whether a proof succeeds in producing conviction that the result it purports to prove is true

---

[3]It seems that only very recently has such a proof of the Law of Cosines been given. See http://www.cut-the-knot.org/pythagoras/CosLawMolokach.shtml (discussed further in Chapter 5 below).

[4]Quoted from Avigad (2006), an article whose concerns overlap to some extent with those of the present text. Avigad suggests that a more fruitful model for analyzing broader aspects of proofs that occur in mathematical practice may be that employed by workers in the field of automated deduction.

depends not only on the formal *correctness* of the argument, but on the mathematical knowledge and sophistication of the audience to which it is presented, as well as that of the mathematical community at large.

The issue here is both a historical and a pedagogical one: On the one hand, standards of rigor have not remained constant, so arguments that once were accepted as convincing by the community of mathematicians of the time may no longer be so regarded; and who is to say that a proof accepted as valid today will not some day be found wanting?[5] On the other hand, a rigorously correct proof may fail to be convincing to those who lack the requisite background or mathematical maturity; and some results (such as the Jordan Curve Theorem) may appear so obvious that mathematical sophistication is required even to understand the *need* for them to be proved.

Recognition of those facts is essential for any meaningful study of the role of alternative proofs in mathematical practice. For to dismiss as proofs arguments that once were, but are no longer, deemed to be correct or complete is to misrepresent mathematical history, by attributing to proofs a permanence they do not possess; and in the present context, it would also eliminate from consideration two of the primary motives for seeking alternative proofs, namely, *correcting errors or filling perceived gaps in previous proofs* (as, e.g., in Hilbert's rigorization of Euclidean geometry), and *presenting arguments that*, though perhaps less rigorous, *are more perspicuous or persuasive to a given audience.*[6]

The primary aim in what follows will be to examine how alternative proofs of various well-known theorems differ. In most cases it will be evident that the proofs *do* differ, and that intuitive feeling can be justified in various ways. For example, proofs that are *direct*, or are *constructive*, may be distinguished from those that are not. A proof that employs a *particular technique* (mathematical induction, for example, or a certain rule of inference) differs tactically from one that does not. One proof may give *greater information* than another — for example, by providing a method for finding a solution to an equation, rather than merely exhibiting one, or by better indicating *why* a result is true. (Showing, e.g., that a convergent real power series has a particular radius of convergence by showing that the corresponding *complex* series has a singularity at a certain point is more informative than simply

---

[5]It is interesting to note, however, how many arguments later deemed to be 'faulty' have yielded correct results — have contained a 'germ' of truth, so to speak. In some cases, the methods originally used to prove such results have been discarded and the theorems reestablished by other, quite different means, while in other instances the original approaches have subsequently been revalidated in light of more sophisticated analyses. (One example is Laurent Schwartz's theory of distributions, which provided a rigorous foundation for arguments based on Dirac's 'δ-function.' Another is Abraham Robinson's creation of non-standard analysis, in terms of which the Newtonian concept of infinitesimal was made comprehensible and arguments based upon it were seen to be correct.)

[6]Proofs may, for example, be crafted to serve the needs of a particular segment within or outside of the mathematical community (students, for example, or lay persons with an interest in mathematics).

applying the ratio test to the given series.) Different proofs may yield *different numerical consequences*, one yielding a better numerical bound, say, than another, or a smaller number that exhibits a certain property. A result employed as a *lemma* in one proof of a theorem may appear as a *corollary* of that theorem if it is proved by other means. One proof of a theorem may be valid in a *wider context* than another. Or one proof of a theorem may be *comprehensible to a particular audience* while another is not.

Proofs may also differ in *how primitive notions are organized into higher-level concepts*, reflecting an *interplay* between *proofs* and *definitions*.

Consider, for example, proving that $\det(AB) = \det(A)\det(B)$. If, for $n \times n$ matrices A, $\det(A)$ is defined as $\sum_{\sigma \in S_n} \mathrm{sgn}(\sigma) \, a_{1\sigma_1} \ldots a_{n\sigma_n}$, where S_n denotes the set of all permutations of $\{1, \ldots, n\}$ and a_{ij} denotes the entry in row i and column j of A, one may compute directly that

$$
\begin{aligned}
\det(AB) &= \sum_{\sigma \in S_n} \mathrm{sgn}(\sigma) \left(\sum_{k_1} a_{1k_1} b_{k_1 \sigma_1} \right) \cdots \left(\sum_{k_n} a_{nk_n} b_{k_n \sigma_n} \right) \\
&= \sum_{k_1,\ldots,k_n} a_{1k_1} \ldots a_{nk_n} \sum_{\sigma \in S_n} \mathrm{sgn}(\sigma) b_{k_1 \sigma_1} \ldots b_{k_n \sigma_n} \\
&= \det(B) \sum_{k \in S_n} \mathrm{sgn}(k) a_{1k_1} \ldots a_{nk_n} \\
&= \det(B)\det(A) = \det(A)\det(B);
\end{aligned}
$$

but the defining formula is complicated and the index chasing even more so.

Instead, with that same definition for $\det(A)$, many linear algebra texts introduce the elementary row operations and the corresponding elementary matrices E, show how each row operation on A results in the matrix EA, and compute easily that $\det(EA) = \det(E)\det(A)$ for any $n \times n$ matrix A and any elementary matrix E. If A is nonsingular, some sequence of m row operations reduces it to the $n \times n$ identity matrix I, so $E_m \ldots E_1 A = I$. Noting that the inverse of any row operation is itself a row operation, $A = E_m^{-1} \ldots E_1^{-1}$, so

$$
\begin{aligned}
\det(AB) &= \det(E_m^{-1} \ldots E_1^{-1} B) \\
&= \det(E_m^{-1}) \ldots \det(E_1^{-1}) \det(B) \\
&= \left(\det(E_m^{-1}) \ldots \det(E_1^{-1}) \right) \det(B) \\
&= \det(A)\det(B).
\end{aligned}
$$

On the other hand, if A is singular, then so is AB, for any $n \times n$ matrix B, so A and AB each reduce to a matrix with one or more rows of zeros, whence $\det(AB) = 0 = \det(A) = \det(A)\det(B)$.

The proof just given is much more perspicuous than the direct one. However, because of its use of inverses, it presumes that the entries in the matrices are elements of a field, whereas the direct proof applies in the wider context of matrices whose entries are elements of a commutative ring.

A third alternative, adopted in texts such as Hoffman and Kunze (1961), is to define a determinant function to be a function from $n \times n$ matrices over a commutative ring R to elements of that ring which is linear as a function of each row of the matrix, which assigns the value 0_R to any matrix having two identical rows, and which assigns the value 1_R to the identity matrix. One then proves that there is exactly one such function, which may be expressed in terms of the entries of the matrix by the aforementioned formula. Without reference to that formula, however, one can show that if D is any function from the $n \times n$ matrices over R to R that is a linear function of each row of any such matrix and that is *alternating* (that is, $D(A) = -D(A')$ whenever A' is obtained from A by interchanging two of its rows), then $D(A) = \det(A)D(I)$. Such a function D is given by $D(A) = \det(AB)$, for *any* fixed $n \times n$ matrix B. So

$$\det(AB) = D(A) = \det(A)D(I) = \det(A)\det(IB) = \det(A)\det(B).$$

In this example, changing the definition of $\det(A)$ from one couched in terms of the entries of the matrix to one that incorporates some of the desired *properties* of the determinant function leads, without loss of generality, to a more abstract proof that is both more perspicuous than the computational one and that explains *why* the formula for $\det(A)$ in terms of the entries of A is what it is.

In comparing different proofs, it must of course be recognized that differences are often a matter of degree. The judgment whether two proofs are essentially different or merely variants of one another (and similar judgments as to whether one is 'simpler', or 'more pure', than another — notions discussed further in the next chapter) is thus a subjective one to a certain extent. Chapter 3 provides a very simple but illustrative example.

The next chapter enumerates some of the reasons why mathematicians have been led to seek alternative proofs of known results. The remaining chapters then provide detailed case studies of particular instances.

References

Aigner, M., Ziegler, G.M.: Proofs from the Book, 2nd ed. Springer, Berlin (2000)

Avigad, J.: Mathematical method and proof. Synthese **193**(1), 105–159 (2006)

Barwise, J.: Mathematical proofs of computer system correctness. Notices Amer. Math. Soc. **36**(7), 844–851 (1989)

Hoffman, K., Kunze, R.: Linear Algebra. Prentice-Hall, Englewood Cliffs, N.J. (1961)

Rav, Y.: Why do we prove theorems? Philosophia Math.(III) **7**, 5–41 (1999)

Chapter 2
Motives for Finding Alternative Proofs

> *Even if we have succeeded in finding a satisfactory solution, we may still be interested in finding another solution. We desire to convince ourselves of the validity of a theoretical result by two different derivations as we desire to perceive a material object through two different senses. Having found a proof, we wish to find another proof as we wish to touch an object after having seen it.*

— George Pólya, *How to Solve It*

Four motives for seeking new proofs of previously established results have already been mentioned in Chapter 1: the desires

(1) **to correct errors or fill perceived gaps in earlier arguments;**
(2) **to eliminate superfluous or controversial hypotheses;**
(3) **to extend a theorem's range of validity; and**
(4) **to make proofs more perspicuous**.

Euclid's efforts to avoid, wherever possible, employing proofs involving superposition of figures exemplify the first of those motives; the persistent attempts, prior to the works of Bolyai and Lobachevsky, to deduce the parallel postulate from the other axioms of Euclidean geometry, the second; Henkin's completeness proof for first-order logic, applicable to uncountable languages as well as to countable ones, the third; and objections, by sixteenth- and seventeenth-century mathematicians, to Archimedean proofs by the method of exhaustion, the fourth.

That Euclid employed the method of superposition when it appeared unavoidable, as in the proof of proposition I,4 (justifying the side-angle-side criterion for congruence), suggests that the ancients considered superposition to be a *perspicuous* principle, but one that rested on *spatial* intuition and so was not rigorously justified by Euclid's (planar) axioms. On the other hand, the *rigor* of Archimedes's proofs by exhaustion was never questioned. But such proofs seemed to many merely to establish *that* a result was true, without providing understanding of *why* it was, or of how the proof might have been discovered.[1] Proofs by mathematical induction are open to similar objections.

---

[1] There is a growing literature on the notion of *explanatory* proofs (those that convey understanding as well as conviction). The article Mancosu (2001) and the book Mancosu (1996) provide useful introductions to that subject.

© Springer International Publishing Switzerland 2015
J.W. Dawson, Jr., *Why Prove it Again?*, DOI 10.1007/978-3-319-17368-9_2

The desire to avoid employing superposition arguments in Euclidean proofs also exemplifies another respect, separate from that of rigor, in which a proof may be deemed deficient: that it fails to exhibit **purity of method**.

Concern for purity of method has arisen frequently in the history of mathematics. Particular instances of such concern include the ancient Greek requirement that geometric constructions be restricted to those performable with straightedge and compass alone; the desires that synthetic proofs be found for results obtained in analytic geometry, that intrinsic proofs be given for theorems in differential geometry or topology, and that proofs of results in model theory not invoke syntactic considerations; the preference for minimizing appeals to analytical or topological methods in proving the Fundamental Theorem of Algebra (discussed further in Chapter 8 below); the quest to find an 'elementary' proof of the Prime Number Theorem[2] (one not employing the methods of analytic number theory), whose unexpected success was among the achievements that led to the award of a Fields Medal to Atle Selberg; and Hilbert's (failed) program to establish the consistency of formalized Peano arithmetic using methods formalizable within that theory itself. In a somewhat broader sense, **concern for methodological propriety** is reflected in such aspects of mathematical practice as the desire to replace indirect or non-constructive proofs by direct or constructive ones, or the debate over whether theorems of analysis ought to be proved by 'soft' (functional-analytic) means (which, though 'slick,' may obscure underlying conceptual motivations) or by 'hard' calculations involving inequalities (which may be lengthy and tedious).

In considering questions of purity, several caveats are in order. First, as remarked in the preceding chapter, in many cases one should more properly speak of *degrees* of purity. Second, it must be recognized that different notions of purity may *conflict*. Consider, for example, Desargues's Theorem in the Plane, which states that if two triangles in the same plane are oriented so that the lines joining corresponding vertices are concurrent, then the corresponding sides, if extended as lines, will intersect in three collinear points. As a theorem of *projective* geometry, one might in the name of purity seek a proof solely by projective means. But as a theorem of *plane* geometry, one might equally well desire a purely planar proof. Those two aims cannot be reconciled, however, since Hilbert showed that any planar proof of that theorem must invoke the *metric* notion of similarity.[3] In addition, it should be noted that some proofs — for example, model-theoretic consistency proofs, in which basic notions are given alternative semantic interpretations — inherently violate purity of method.

Proofs that violate purity of method may, however, possess merits of their own. The example given in the previous chapter of determining the radius of convergence

---

[2]In its simplest form, the Prime Number Theorem (the subject of Chapter 10 below) states that $\lim_{x \to \infty} \pi(x) \left/ \dfrac{x}{\ln x} \right. = 1$, where $\pi(x)$ denotes the number of primes less than x.

[3]See Chapter 9 below for a detailed discussion of Desargues's Theorem. An illuminating discussion of Hilbert's proof is given on pp. 222–229 of Hallett (2008).

of a real power series is a case in point: consideration of the singularities of the corresponding complex series makes it clear *why* the radius of convergence is what it is — an insight that could not be obtained without introducing the complex perspective. The more general context here has greater explanatory power.

Concern for purity of method implies a restriction on means of proof, but not conversely. Other reasons for restricting allowable means of proof include **reconstructing proofs employed in antiquity**, given what we know about the state of mathematical knowledge in particular ancient cultures,[4] and **benchmarking** (demonstrating the power of a given methodology by employing it to prove theorems in areas where it might seem not to be applicable).[5]

Methodologically 'pure' proofs demand fewer conceptual prerequisites for their understanding, since they employ no notions beyond those implicit in the statement of the theorem to be proved. They may, however, be long and complex (as are elementary proofs of the Prime Number Theorem), and thus lack **elegance**, an aesthetic characteristic that is hard to define but is nonetheless readily perceived and highly esteemed by mathematicians.

Proofs that are elegant may employ sophisticated concepts, but they are usually short, often employ novel perspectives or strategies, and generally convey immediate understanding and conviction (producing an *Aha!* reaction). Reading such proofs yields deep intellectual enjoyment and satisfaction, akin to that experienced in viewing fine works of art or listening to great music. Elegant proofs are, however, often of limited generality, involving insights applicable only to a particular problem.

Another aesthetic criterion according to which proofs may be compared is **simplicity**. One proof may be simpler than another in various respects. For example:

(1) It may be significantly shorter.
(2) It may involve fewer conceptual prerequisites.
(3) It may reduce the extent of computations to be performed or the number of cases to be considered.

A well-known example of (3) is Hilbert's basis theorem for invariants of algebraic forms,[6] whose non-constructive proof swept away in one stroke a tangle of laborious calculations in invariant theory, including much of the life work of the mathematician Paul Gordan.

---

[4]One example is discussed in Chapter 4.

[5]Such as using topological arguments to prove results in mathematical logic. Another example is Errett Bishop's text *Foundations of Constructive Analysis* (Bishop 1967), which Bishop himself called "a piece of constructivist propaganda," written to demonstrate how large a part of abstract analysis can be developed within a constructive framework.

[6]In the form proved by Hilbert, the theorem states that every ideal in the ring of multivariate polynomials over a field is finitely generated, so that for any set of polynomial equations, there is a finite set of such equations that has the same set of solutions.

Remarkably, Hilbert himself believed that among all proofs of a theorem there must always be one that is simp*lest*. He said so explicitly in the statement of what he had intended would be the twenty-fourth problem in the list he drew up for presentation in his famous address at the Second International Congress of Mathematicians. Due to time constraints, however, he mentioned only ten of the problems during the lecture itself. Thirteen more appeared in the version of his address published in the Conference *Proceedings*, but the twenty-fourth problem came to light only in the mid-1990s, when it was discovered in one of the notebooks in Hilbert's *Nachlass*.[7] It asked for "Criteria of simplicity, or proof of the greatest simplicity of certain proofs," with the understanding that "under a given set of conditions there can be but one simplest proof." Hilbert never posed the problem in public, perhaps because of the difficulty of making the notion of "simplicity" formally precise.[8] Accordingly, judgments of whether one proof is simpler than another have up to now been based primarily on informal criteria like those above.

In addition to the practical and aesthetic rationales so far considered for presenting new proofs of previously established theorems, there are also more personal motives for doing so. For mathematics is, after all, a *human* endeavor. Skill in proving theorems is best developed by attempting to prove results on one's own, and in the course of doing so, one may well devise an argument not previously given, since people do not all think alike. In some cases, one may not be *aware* of other proofs that have been given—different proofs, e.g., may arise in different cultures, or a new result may be discovered, simultaneously and independently, by different individuals using different arguments. Like all fields of scholarship, mathematics is also a competitive enterprise, and having seen a proof presented by someone else, one may be challenged to devise one's own proof of it. Thus, apart from the reasons already enumerated, alternative proofs may arise simply as **expressions of individual patterns of thought**, perhaps reflecting personal predilections or preferences for using particular tools.

Here an analogy may be made between mathematics and the sport of mountaineering. Mathematicians are driven to solve problems for the same reason that mountaineers are driven to climb mountains: because they are there; and as in mountaineering, **pioneering a new route** to a summit, perhaps using restricted means, may be just as challenging and exciting (and be accorded just as much respect by one's peers) as being the first to make the ascent.[9] Mascheroni's work showing that the compass alone suffices to carry out all straightedge and compass constructions may, for example, be compared with ascents by climbers who disdain

---

[7]"Mathematisches Notizbuch" (Cod. ms. D. Hilbert 600), preserved in the Handschriftenabteilung of the Niedersächische Staats- und Universitätsbibliothek, Göttingen.

[8]See Thiele and Wos (2002) for further details on the history of the twenty-fourth problem and on results related to it found recently by those working in automated theorem proving.

[9]Jon Krakauer, e.g., in his book *Into Thin Air*, wrote: "Getting to the top of any given mountain was considered much less important than *how* one got there: prestige was earned by tackling the most unforgiving routes with minimal equipment, in the boldest style imaginable."

supplemental oxygen or mechanical aids. Some new proofs may thus be presented purely as such, especially if they are deemed to involve particularly clever or unusual insights, or to exhibit particular **economy of means**.

Apart from such specific motives for giving alternative proofs, there is also an over-arching purpose, often overlooked, that multiple proofs serve — one analogous to the role of *confirmation* in the natural sciences. Namely, just as agreement among the results of different experiments heightens credence in scientific hypotheses (and so also in the larger theories within which those hypotheses are framed), different proofs of theorems bolster confidence not only in the particular results so proved, but in the overall structure and coherence of mathematics itself. As Max Dehn remarked in an address delivered 18 January 1928, "Most [mathematical] results are so involved in the general web of theorems, they can be reached in so many ways, that their incorrectness is simply unthinkable." (Dehn 1983, p. 22) That is, trust in mathematical results is based rather on the "multitude and variety" of the deductions that lead to them than on "the conclusiveness of any one" of those deductions. Mathematical reasoning is not "a chain, ... no stronger than its weakest link, but a cable," whose fibers, though "ever so slender," are "numerous and intimately connected."[10]

References

Bishop, E.: Foundations of Constructive Analysis. McGraw-Hill, New York (1967)

Dehn, M.: The mentality of the mathematician. A characterization. Math. Intelligencer **5**, 18–26 (1983)

Hallett, M.: Reflections on the purity of method in Hilbert's *Grundlagen der Geometrie*. In Mancosu, P., The Philosophy of Mathematical Practice, pp. 198–255. Oxford U.P., Oxford (2008)

Mancosu, P.: Philosophy of Mathematics and Mathematical Practice. Oxford U.P., New York and Oxford (1996)

Mancosu, P.: Mathematical explanation—problems and prospects. Topoi **20**(1), 97–117 (2001)

Pierce, C.S.: Some consequences of four incapacities. J. Speculative Philos. **2**, 140–157 (1868)

Thiele, C., Wos, L.: Hilbert's twenty-fourth problem. J. Automated Reasoning **29**(1), 67–89 (2002)

---

[10]To paraphrase a remark C.S. Peirce made in regard to philosophy (Pierce 1868).

Chapter 3
Sums of Integers

As a first, very simple case study of alternative proofs, consider the following four proofs of the identity $1 + 3 + \cdots + (2n - 1) = n^2$ (a result known since antiquity, and one that is readily conjectured from specific instances):

1. **Proof via gnomons** (Figure 3.1)
2. **Proof by induction**:
   The result is true when $n = 1$, since $1 = 1^2$. Assuming that it is true when $n = k$, suppose $n = k + 1$. Then $1 + 3 + \cdots + (2k - 1) + (2(k + 1) - 1) = 1 + 3 + \cdots + (2k + 1) = k^2 + (2k + 1) = (k + 1)^2$. By induction, the result therefore holds for all n.
3. **Proof via Gauss's method** (recalling the tale of his schoolboy summation of the first hundred integers[1]):

$$1 + 3 + 5 + \cdots + (2n - 1) = S$$
$$(2n - 1) + \cdots + 5 + 3 + 1 = S$$

   whence, adding the equations together column by column, $2n + 2n + \cdots + 2n = 2S$, where the left member of the equation has n equal summands. Thus $2n \cdot n = 2S$, or $S = n^2$.
4. **The stairstep proof** (Figure 3.2)
   Which of these proofs are to be regarded as essentially different, and on what grounds?

---

[1]For an interesting commentary on the origin and evolution of that possibly apocryphal tale, see Hayes (2006).

© Springer International Publishing Switzerland 2015
J.W. Dawson, Jr., *Why Prove it Again?*, DOI 10.1007/978-3-319-17368-9_3

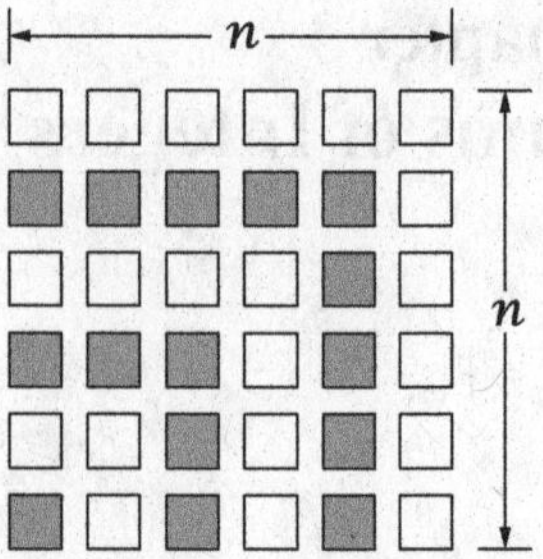

Fig. 3.1 The addends form gnomons

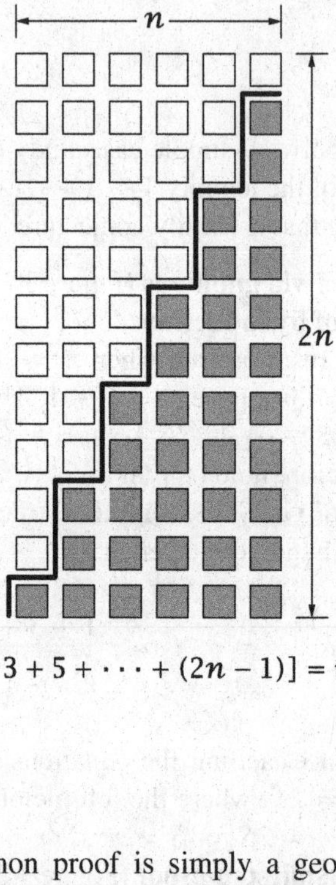

$$1 + 3 + 5 + \cdots + (2n - 1) = n^2$$

Fig. 3.2 The addends form a staircase

$$2[1 + 3 + 5 + \cdots + (2n - 1)] = n \cdot 2n$$

3.1   Comparative analysis

At first glance, one might think that the gnomon proof is simply a geometric representation of the inductive one. In several respects, however, it is not. In particular, it is *more perspicuous* and requires *fewer prerequisites* for its understanding: Given a square with sides of any integral length, it is clear that to obtain a square with sides one unit larger, it suffices to add unit squares along two adjacent edges of the given square and then place one unit square in the corner between those edges. It follows that if the original square has side length n, a gnomon composed of $2n + 1$ unit squares must be added to form a square of side length $n + 1$. Starting with a single unit square and successively adjoining such gnomons thus corresponds to starting with the number 1 and adding successive odd numbers. The only prerequisites to understanding the proof are knowing what an odd number and a geometric square are, and how the area of a square is found.[2] No knowledge whatever of algebra is needed, in contrast to the rudimentary algebra employed in the inductive proof.

In addition, arranging counters in the shape of gnomons[3] readily suggests *what* the sum of the first n successive odd numbers should *be* — a conjecture that must be made before an inductive argument can be carried out. And as teachers of the subject will readily attest, the very *concept* of mathematical induction is a difficult one for many students to grasp.

For all these reasons, the gnomon and inductive proofs must be regarded as distinct. But what of the other two proofs?

Both Gauss's method (as described above[4]) and the stairstep proof involve *duplication* and *inversion*, of an equation in one case and a figure in the other. Both also seem equally perspicuous, since the recognition that the area of the $n \times (n + 1)$ rectangle in the stairstep proof is given by $n(n + 1)$ involves essentially the same insight as recognizing that that product gives the sum of n summands each equal to $n + 1$. So it *does* seem reasonable to regard the stairstep proof merely as a geometric representation of the Gaussian one.

That the Gaussian/stairstep proof is conceptually distinct from the other two proofs also seems clear. The diagrams that constitute the gnomon and stairstep proofs represent two different (both quite natural) ways of arranging successive odd numbers of counters into patterns. One yields a square while the other yields an $n \times (n + 1)$ rectangle. One gives the result directly, the other requires division by two. And no induction is involved in the Gaussian argument.

On those grounds alone, then, we may conclude that three of the four proofs considered above are essentially different. To what extent can each of them be generalized?

The method of mathematical induction is, of course, a very general one. It can be employed in a great variety of contexts, once a result has been conjectured by other means (including *ordinary* induction). It is not a method for discovering results, nor (in general) for providing explanatory proofs; but it can be useful for validating a result that is needed, when other means of proving it would require that a substantial amount of background theory be developed.

---

[2] No wonder, then, that the gnomon proof has very ancient origins. See in particular the commentary by Sir Thomas Heath on pp. 358–360 of vol. I of his translation of Euclid's *Elements* (Heath 1956).

[3] The term originated in the ancient Greek theory of figured numbers, where it meant "a number which, when added to a term in a given class of figured numbers, produces the next number in that class" See Knorr (1975), pp. 135–154, for further examples and discussion.

[4] This qualification is necessary, because as Brian Hayes points out in the article cited in footnote 1 above, not all accounts of Gauss's schoolboy triumph describe that same algorithm. Several, e.g., say that he folded the series $1 + 2 + 3 + \cdots + 100$ in the middle and added corresponding entries in the two rows — a method that yields the same numerical result without having to divide by two, but which applies only to series with an *even* number of terms, and so does not yield the general formula for an arbitrary number of summands.

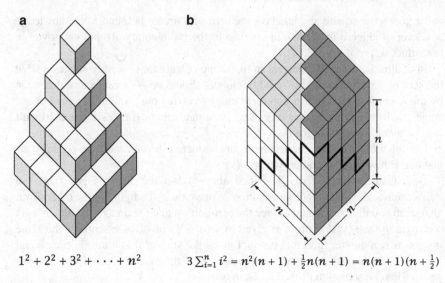

$$1^2 + 2^2 + 3^2 + \cdots + n^2 \qquad 3\sum_{i=1}^{n} i^2 = n^2(n+1) + \tfrac{1}{2}n(n+1) = n(n+1)(n+\tfrac{1}{2})$$

Fig. 3.3 Stacked pyramids

The Gaussian/stairstep technique can obviously be applied to find the sum of any finite *arithmetic* series — in particular, to show that

$$\sum_{i=1}^{n} i = \frac{n(n+1)}{2}.$$

With sufficient powers of three-dimensional visualization it can also be employed to find $\sum_{i=1}^{n} i^2$: Form a stepped pyramid by stacking square layers of sizes $n \times n, (n-1) \times (n-1), \ldots, (2 \times 2), 1 \times 1$ atop one another, aligning all their "back" corners (Figure 3.3(a)). Then note that three such stepped pyramids can be fitted together to form a solid composed of an $n \times (n+1)$ rectangular parallelepiped topped by half of an $n \times (n+1)$ layer of unit cubes (Figure 3.3(b)).[5] Hence

$$3\sum_{i=1}^{n} i^2 = n(n+1)(n+\tfrac{1}{2}).$$

Likewise, the gnomon method is readily modified to find the sum of the first n even numbers, by starting with two counters and using gnomons to form successive rectangles of dimensions $n \times (n+1)$, for each n (what the ancient Greeks called the "oblong" numbers). Half that sum then again gives

---

[5]Figure 3.3 is based on that in Nelsen (1993), p. 77, where the construction is credited to Man-Keung Siu. ©The Mathematical Association of America 2013. All right reserved.

$$\sum_{i=1}^{n} i = \frac{n(n+1)}{2}.$$

Gnomons, too, can be applied in three dimensions to find the equation

$$(n+1)^3 - n^3 = 3n^2 + 3n + 1,$$

the three-dimensional analogue of the equation $(n+1)^2 - n^2 = 2n+1$; and if those two equations are added together, the resulting expression for $(n+1)^3$ in terms of lower powers of n can be used, in conjunction with the formula for $\sum_{i=1}^{n} i$, to provide an alternate derivation of $\sum_{i=1}^{n} i^2$.

To carry out the construction, form an $(n+1) \times (n+1) \times (n+1)$ cube from a given $n \times n \times n$ cube by placing a gnomon of $2n+1$ unit cubes around the base of the latter (Figure 3.4(a)). Then stack n more of those gnomons atop the first one (Figure 3.4(b)), and add a final $n \times n$ square layer of unit cubes above the original cube. The resulting "gnomonic" shell around the given cube will thus be made up

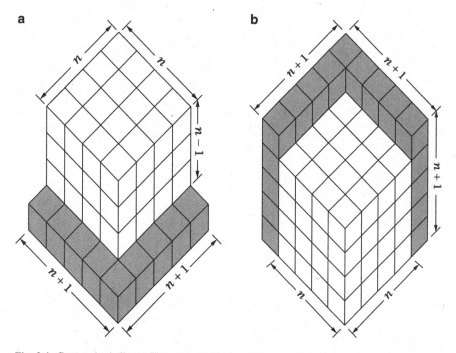

Fig. 3.4 Gnomonic shells. (a) First step. (b) Next-to-last step

of $n + 1$ gnomons, each containing $2n + 1$ unit cubes, plus a square "roof" of n^2 unit cubes atop the starting cube. So

$$(n + 1)^3 - n^3 = (n + 1)(2n + 1) + n^2 = 3n^2 + 3n + 1.$$

These examples illustrate how the gnomon and stairstep methods can be used to establish many of the same results via different routes.

References

Hayes, B.: Gauss's day of reckoning. Amer. Sci. **94**(3), 200–205 (2006)

Heath, T.: The Thirteen Books of Euclid's Elements (3 vols.). Dover, New York (1956)

Knorr, W.: The Evolution of the Euclidean Elements. A Study of the Theory of Incommensurable Magnitudes and its Significance for Early Greek Geometry. Reidel, Dordrecht (1975)

Nelsen, R.B.: Proofs without Words. Exercises in Visual Thinking. Math. Assn. Amer., Washington, D.C. (1993)

Chapter 4
Quadratic Surds

This chapter provides an example of how an alternative proof may be used to provide a rational reconstruction of a historical practice. It concerns the following well-known

Theorem: $\sqrt{n}$ *is rational if and only if it is integral, that is, if and only if n is a perfect square.*

That result is an almost immediate consequence of

Euclid's proposition VII,20, which (in modern terminology) states: *If a, b, c, d are natural numbers and a is the least natural number for which $a/b = c/d$ (that is, a/b is in lowest terms), then a divides c and b divides d.*[1]

For suppose $\sqrt{n} = p/q$, where p and q are relatively prime. Then $p/q = nq/p$. By proposition VII,20, q must divide p, whence $q = 1$.

Note that this proof is a direct one, and the value of n plays no role in it. That argument, however, was apparently not known at the time that the irrationality of many such quadratic surds was first discovered. Evidence that it was not is provided by an intriguing passage in Plato's dialogue *Theaetetus*, cited by Sir Thomas Heath in his introductory commentary to Book X of Euclid's *Elements* (Heath 1956, vol. III, p. 1–2). *Theaetetus* concerns the pre-Euclidean mathematician of that name, who lived from about 415 to 368 BCE; and in the passage cited, Theaetetus says that his teacher, Theodorus of Cyrene, "was proving for us via diagrams something about roots, such as the roots of three or five, showing that they are incommensurable by the unit. He selected other examples up to seventeen, where, for some reason, he encountered difficulty." The passage is vague concerning just how Theodorus established the irrationality of $\sqrt{3}$, $\sqrt{5}$, etc., but it certainly suggests (as Heath goes

---

[1] As noted already in Zeuthen (1896), pp. 156–157, Euclid's proof of that proposition is faulty. Proposition VII,20 does, however, follow by *reductio* from the division algorithm and proposition VII,17 (that $b/c = ab/ac$ for any natural numbers a, b, c).

© Springer International Publishing Switzerland 2015
J.W. Dawson, Jr., *Why Prove it Again?*, DOI 10.1007/978-3-319-17368-9_4

Fig. 4.1 A triangular tiling

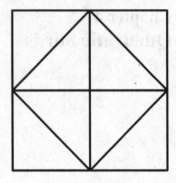

on to note) that *separate* proofs were involved, which would not have been necessary had Theodorus known the general proof given above.

Heath further notes that the passage makes no mention of Theodorus's having proved that $\sqrt{2}$ is incommensurable with the unit, "doubtless for the reason that its incommensurability had been proved before." As to when and how that was done, Heath speculates that even *without* knowledge of the Pythagorean Theorem, the ancient Greeks could easily have deduced from a diagram like that in Figure 4.1 that the hypotenuse of an isosceles right triangle whose legs are of unit length must have length $\sqrt{2}$ (Heath 1956, vol. I, p. 352).

He then refers to Aristotle's *Prior Analytics* (I,23) for an explanation of how the discovery that $\sqrt{2}$ is incommensurable with the unit was made. According to Aristotle,

> [A]ll who effect an argument *per impossibile* infer syllogistically what is false and prove the original conclusion hypothetically when something impossible results from the assumption of its contradictory; e.g., that the diagonal of the square is incommensurable with the side, because odd numbers are equal to evens if it is supposed to be commensurate.

That is, the original proof that $\sqrt{2}$ is irrational was an indirect one, based on *parity* considerations. It presumably went as follows (a proof still often presented in classrooms today):

Suppose that $\sqrt{2}$ is equal to the ratio of two integers, p and q. After canceling any common factors, p and q may be assumed to be relatively prime, so one of them must be odd. Now $p^2 = 2q^2$, so p^2 is even. Hence p itself must be even, say $p = 2m$. But then $4m^2 = 2q^2$, or $2m^2 = q^2$, so q must also be even, contradiction. Hence $\sqrt{2}$ must be irrational.

Aristotle's text says nothing, however, about the incommensurability of lengths representing square roots of other integers. So the questions remain: How did Theodorus obtain the results Theaetetus ascribed to him, and what problem arose with $\sqrt{17}$?

A plausible answer to both those questions was first suggested by Jean Itard in his book (Itard 1961). He pointed out that proofs based on parity considerations could be used to establish the irrationality of each of the roots mentioned by Theaetetus, but — tellingly — *not* the irrationality of $\sqrt{17}$. With the benefit of elementary algebra, Itard's claim can be justified as follows:

Suppose that n is the *least* non-square positive integer whose square root is rational, say $\sqrt{n} = p/q$, where p and q are relatively prime. Then $p^2 = nq^2$, and q cannot be even, since if it were, p would be also. So q is odd. Then if p were even, n would be also, say $p = 2k$ and $n = 2j$. Replacing p and q by those expressions yields $4k^2 = 2jq^2$, or $2k^2 = jq^2$. Thus j would have to be even, say $j = 2r$. Hence $2k^2 = 2rq^2$, or $k^2 = rq^2$, an equation which has the same form as $p^2 = nq^2$. So $\sqrt{r} = k/q$, with $r < 4r = n$; and if r were a perfect square, n would be too, contrary to assumption. Thus r is *not* a perfect square, but is a positive integer less than n with a rational square root; and that contradicts the minimality of n. The assumption that p could be even is thus untenable. Consequently, both p and q must be odd. Therefore so are p^2 and q^2, which means n must be odd as well. Writing $p = 2k + 1, q = 2j + 1$ and $n = 2m + 1$, the equation $p^2 = nq^2$ becomes

$$(2k + 1)^2 = n(2j + 1)^2,$$

that is,

$$4k^2 + 4k + 1 = n(4j^2 + 4j + 1) = 4j^2 n + 4jn + n,$$

which may be rearranged as

$$4(k^2 + k - j^2 n - jn) = 4[k(k + 1) - nj(j + 1)] = n - 1 = 2m.$$

Since the product of any two *consecutive* positive integers must be even, it follows that the bracketed expression in the last equation above is an even number, say $2s$, so $8s = 2m$. Hence, finally, $n = 2m + 1 = 8s + 1$.

Conclusion: If there is an integer n, not a perfect square, whose square root is rational, then the least such n must be of the form $8s + 1$ (that is, be congruent to 1 mod 8); and the least non-square integer of that form is 17.

Of course, the ancient Greeks could not have carried out a general algebraic analysis like that just given. But Wilbur Knorr, in chapter VI of his book (Knorr 1975), showed how, in each individual case, geometric methods available to the ancient Greeks could have been used to establish the parity-theoretic results needed to prove that $\sqrt{n}$ is irrational, for every non-square integer from $n = 2$ to $n = 15$. He further showed that those methods would break down in the case $n = 17$, thus justifying Theaetetus's account of Theodorus's works.

Could the geometric methods of the ancient Greeks have been employed to show that $\sqrt{n}$ is irrational for *any* integer n that is not a perfect square? The following *reductio* proof by infinite descent, recently devised by Akihiro Kanamori, gives an affirmative answer.

Kanamori's geometric proof: Suppose that $\sqrt{n} = a/b$, where a and b have no common factor and $b \neq 1$. By the division algorithm, $a = bq + r$ for some integers q and r, with $0 < r < b$. Construct a right triangle with hypotenuse of length a and one leg of length bq, and let c denote the length of the other leg. Then draw an

Fig. 4.2 A simple
reductio proof

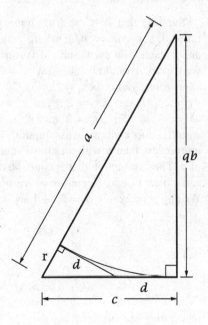

arc of radius bq from the vertex at the right angle to the hypotenuse, and from the
point on the hypotenuse so determined, construct the perpendicular segment lying
inside the triangle (Figure 4.2). Call the length of that segment d, and note that
the segment from the vertex at the right angle to the point where the perpendicular
segment intersects the leg of length c must also be d, since those segments are
tangents to a circle from a common point. The right triangle with legs of length r
and d is then similar to the original triangle, so $bq/c = d/r$ and $a/c = (c - d)/r$.
Therefore $cd = bqr$ and $a = (c^2 - cd)/r$. But also, by the Pythagorean theorem,
$c^2 = a^2 - b^2q^2 = b^2n - b^2q^2$, since $a = b\sqrt{n}$. Hence

$$\sqrt{n} = \frac{a}{b} = \frac{b^2n - b^2q^2 - bqr}{br} = \frac{bn - q(bq + r)}{r} = \frac{bn - qa}{r}.$$

Thus $\sqrt{n}$ has been represented as a ratio of two integers, with denominator $r < b$.
But that is absurd, since the construction could be repeated indefinitely to yield a
never-ending strictly decreasing sequence of positive integers less than r.

Note that the proof just given may be regarded as a geometrization of the one-line
algebraic proof

$$\frac{a}{b} = \frac{a\frac{r}{b}}{r} = \frac{a(\sqrt{n} - q)}{r} = \frac{a\sqrt{n} - aq}{r} = \frac{bn - aq}{r}.$$

References

Heath, T.: The Thirteen Books of Euclid's Elements (3 vols.). Dover, New York (1956)

Itard, J.: Les livres arithmétique d'Euclide. Hermann, Paris (1961)

Knorr, W.: The Evolution of the Euclidean Elements. A Study of the Theory of Incommensurable Magnitudes and its Significance for Early Greek Geometry. Reidel, Dordrecht (1975)

Zeuthen, H: Geschichte der Mathematik im Altertum und Mittelalter. Andr. Fred. Höst & Sön, Copenhagen (1896)

Chapter 5
The Pythagorean Theorem

The Pythagorean Theorem is one of the oldest, best known, and most useful theorems in all of mathematics, and it has also surely been proved in more different ways than any other. Euclid gave two proofs of it in the *Elements*, as Proposition I,47, and also as Proposition VI,31, a more general but less well-known formulation concerning arbitrary 'figures' described on the sides of a right triangle. The first of those demonstrations is based on a comparison of areas and the second on similarity theory, a basic distinction that can be used as a first step in classifying many other proofs of the theorem as well.

Since Euclid's time hundreds of other proofs have been given — not because the correctness of the result or the rigor of Euclid's arguments have ever been questioned, but principally because the theorem has fascinated generations of individuals, not only professional mathematicians but students and amateurs, who have felt challenged to apply their own ingenuity to prove it. The multitude of proofs thus created stands as an exemplar *par excellence* of the desire to find a previously undiscovered path to a goal.

Extensive compilations of proofs of the Pythagorean Theorem have appeared in several publications. Sources in English include a series of twelve articles by Benjamin F. Yanney and James A. Calderhead (Yanney and Calderhead 1896–9) that appeared in vols. 3 through 6 of the *American Mathematical Monthly*, each entitled "New and old proofs of the Pythagorean theorem"; the book (Loomis 1940) by Elisha S. Loomis, first published in 1927 and reprinted in 1968 by the National Council of Teachers of Mathematics; and the geometry web pages maintained by Alexander Bogomolny (Bogomolny 2012).

The first of those references nominally presents 100 different proofs; the second, 367; and the third, 96; but the qualifier 'nominally' is important for several reasons. First, as each of the compilers points out, some of the proofs admit numerous variations (sometimes thousands in the case of similarity arguments, depending on which particular sets of proportions are employed). Second, especially in Loomis's book, distinctions among proofs are often not clearly or carefully made, despite his proclaimed intent to classify and arrange the proofs according to "method of

© Springer International Publishing Switzerland 2015
J.W. Dawson, Jr., *Why Prove it Again?*, DOI 10.1007/978-3-319-17368-9_5

proof and type of figure used." Indeed, the very first of his 'algebraic' proofs
— those based on similarity relations, as opposed to 'geometric' proofs based
on area comparisons — appears to differ little, if at all, from his 'algebraic'
proofs 38 and 93 (described as involving "the mean proportional principle" and
"the theory of limits," respectively), or from 'geometric' proof 230; and it is also
similar to Euclid's Proposition VI,31 — of which, incredibly, Loomis appears to
have been completely unaware! Third, not all of the proofs given in Yanney's
and Calderhead's articles or in Loomis's book are *correct*! The most egregious
example is Loomis's 'algebraic' proof 16, apparently taken over uncritically from
Yanney's and Calderhead's proof X. Ostensibly a proof by *reductio*, it assumes
the Pythagorean Theorem to be true, derives a *true* consequence from it, and then
declares the assumption to have been justified![1] Several other fallacious proofs from
those two earlier sources are also cited on Bogomolny's web site.

Loomis's book is problematical on other grounds as well. The very idea, for
example, of distinguishing proofs according to the diagrams used to represent them
seems fatally flawed, both because the *same* diagram, interpreted differently, may
be used to represent conceptually distinct arguments, and because, conversely, some
arguments can be represented by *more than one* distinct diagram. (See below for
further discussion of both points.) Loomis's criteria for excluding some proofs
on the grounds that they are mere variants of ones given, while including others
that seem hardly distinguishable from them, are vague, to say the least; and in
several instances he also made categorical declarations with no, or with only weak,
attempts at justification. Thus, for example, he declared the first of the 'algebraic'
proofs he listed (the third of the proofs considered below) to be "the shortest
proof possible," without further discussion.[2] He also claimed (pp. viii and 224)
that "no trigonometric proof [of the Pythagorean Theorem] is possible," because
"all the fundamental formulae of trigonometry are themselves based upon [the
trigonometric form of] that theorem," namely, the identity $\cos^2 \theta + \sin^2 \theta = 1$.
But that is simply false: A very simple derivation of that identity, based directly on
the ratio definitions for sine and cosine, is given below (proof **4**).[3]

The purpose of the analyses that follow is not to duplicate the excellent
commentaries on Bogomolny's site (though all the proofs considered below are to
be found there), nor to provide a critique of all, or even most, of the proofs collected
by Yanney and Calderhead or Loomis. Rather, the aim of the case study undertaken
in this chapter is the more modest one of examining and comparing seven proofs of

---

[1] It is reprehensible that such a blatantly invalid proof was reproduced without comment in the
NCTM reprint of Loomis's book.

[2] Eli Maor, for one, has disagreed. In his book (Maor 2007) he proposes another candidate and
provides much other illuminating discussion of the Pythagorean Theorem and various proofs
thereof.

[3] Alternatively, in Zimba (2009) it is shown that the Pythagorean Theorem follows easily from the
identities for $\sin(\alpha - \beta)$ and $\cos(\alpha - \beta)$, each of which, as is well known, can be derived directly
from the ratio definitions (see, e.g., Nelsen 2000, pp. 40 and 46).

Fig. 5.1 Euclid's 'windmill' diagram

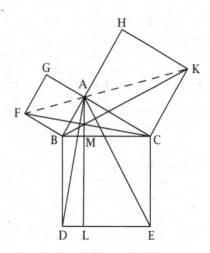

the Pythagorean Theorem, selected as representative examples of distinct conceptual approaches that have been taken over the centuries. The two proofs given by Euclid[4] provide a natural starting point.

1. *Elements*, **Book I, Proposition 47:** In right-angled triangles, the [area of the] square on the side subtending the right angle is equal to [the sum of the areas of] the squares on the sides containing the right angle.

Proof summary: The proof is based on the well-known 'windmill' or 'bride's chair' diagram (Figure 5.1), constructed as follows:

Given the right triangle *ABC* with right angle *BAC*, erect squares on each of its sides (as justified by the immediately preceding proposition I,46), and note that since angles *BAC, BAG,* and *CAH* are all right angles, $\overline{AH}$ and $\overline{AG}$ are extensions of the segments $\overline{BA}$ and $\overline{CA}$, so that segments $\overline{CG}$ and $\overline{BH}$ are parallel to segments $\overline{BF}$ and $\overline{CK}$, respectively. Next, drop a perpendicular from *A* to segment $\overline{DE}$, intersecting $\overline{BC}$ at *M* and $\overline{DE}$ at *L*. The segment $\overline{LM}$ then divides the square on $\overline{BC}$ into two rectangles, and *the central idea of the proof is to show that the area of the square on $\overline{AB}$ is equal to the area of the rectangle BDLM and the area of the square on $\overline{AC}$ is equal to the area of the rectangle CELM*. The principal tool for doing so is Proposition I,41, which states, in essence, that all triangles with the same base and altitude have the same area, or in other words, that *the area of a triangle is invariant under the action of any shear transformation parallel to its base*. Accordingly, if the segments $\overline{AD}, \overline{AE}, \overline{CF},$ and $\overline{BK}$ are drawn, then triangles *BCF* and *ABF* have the same area (half the area of the square on $\overline{AB}$), as do triangles *BCK* and *ACK* (half the area of the square on $\overline{AC}$); and likewise, triangle *ABD* has half the area of rectangle *BDLM* and triangle *ACE* half the area of rectangle *CELM*. But triangles *BCF* and *ABD* are congruent, as are triangles *BCK* and *ACE*, by the side-angle-side

---

[4]Both due to Euclid himself, according to Proclus.

criterion (Proposition I,4), since angles *CBF* and *DBA* are equal (each being a right angle plus angle *ABC*), as are angles *BCK* and *ACE* (each being a right angle plus angle *ACB*). Hence half the area of the square on $\overline{AB}$ plus half the area of the square on $\overline{AC}$ is equal to half the area of rectangle *BDLM* plus half the area of rectangle *CELM*, from which the desired result follows by multiplying by two.

2. *Elements*, **Book VI, Proposition 31:** In right-angled triangles the figure on the side subtending the right angle is equal to the similar and similarly described figures on the sides containing the right angle. (See Figure 5.2.)

   Overview of proof: Although its statement is much more general than that of Proposition I,47, the proof of Proposition VI,31 is much simpler. Indeed, the proof follows by inspection from Figure 5.3 below, once the following facts are recognized:

   (a) The statement of the proposition does not require that the figures described on the sides of the triangle must be *exterior* to the triangle, nor that they not overlap.
   (b) Similar figures differ only in *scale*.
   (c) If two figures are *rescaled* by the same factor, the *ratio* of their areas is unchanged.

   For in Figure 5.3 we may regard triangle *ABD* as described on side $\overline{AB}$ of triangle *ABC*, triangle *ACD* as described on side $\overline{AC}$, and triangle *ABC* as described on its own hypotenuse $\overline{BC}$. (Alternatively, we may consider Figure 5.4, in which the triangles *ABD*, *ACD*, and *ABC* of Figure 5.3 have been reflected around the lines

Fig. 5.2 Euclid's
Proposition VI,31

Fig. 5.3 Three similar
triangles

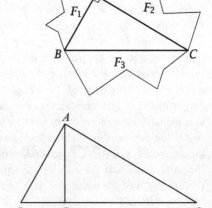

Fig. 5.4 Figure 5.3
"unfolded"

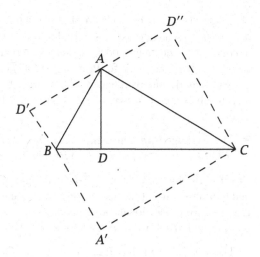

$\overline{AB}$, $\overline{AC}$ and $\overline{BC}$ to form the triangles ABD', ACD'', and $A'BC$.[5]) Triangles ABD, ACD, and ABC are similar because their corresponding angles are equal, and since triangles ABD and ACD exactly fill up triangle ABC,

$$\text{area of } ABD + \text{area of } ACD = \text{area of } ABC.$$

The proposition therefore holds for those triangles. But the general case then follows from (b) and (c), since F_1, F_2, and F_3 are similar by hypothesis and are scaled, like triangles ABD, ACD, and ABC, in proportion to the lengths of $\overline{AB}$, $\overline{AC}$ and $\overline{AD}$. So, for some constant k, we have

$$\frac{\text{area of } F_1}{\text{area of } ABD} = \frac{\text{area of } F_2}{\text{area of } ACD} = \frac{\text{area of } F_3}{\text{area of } ABC} = k.$$

Multiplying the first displayed equation above by k then gives

$$\text{area of } F_1 + \text{area of } F_2 = \text{area of } F_3, \qquad\qquad \text{q.e.d.}$$

The generality of Proposition VI,31, coupled with the economy of means used to prove it, is breathtaking. So why did Euclid give the less general Proposition I,47 with its more involved proof? Presumably, as Heath says (Heath 1956, vol. I, p. 355), because in the plan of exposition that Euclid adopted for the *Elements*, the development of Eudoxus's ingenious theory of proportions (needed in order for similarity theory to apply to incommensurable as well as commensurable magnitudes) was postponed to Book V, whereas the more restricted form of the Pythagorean Theorem given in I,47 was needed early on. Alternatively, as Heath

---

[5]In Maor (2007) the argument based on Figure 5.4 is called the 'folding bag' proof.

also suggests, it may be that a proof of the Pythagorean Theorem based on similarity theory was first advanced *prior* to Eudoxus's work, but was recognized to apply only to *commensurable* magnitudes, so that an alternative proof independent of similarity theory (such as that given for Proposition I,47) was desired and subsequently found.

Note that, despite its reliance on similarity theory, the proof of Proposition VI,31 *also* involves a comparison of areas, in contrast to the next proof (one commonly given today):

3. The Pythagorean Theorem: *Given any right triangle whose hypotenuse* $a = |\overline{BC}|$, *let* $b = |\overline{AC}|$ *and* $c = |\overline{AB}|$. *Then* $a^2 = b^2 + c^2$.

Proof: In Figure 5.3, let $y = |\overline{BD}|$. Triangle ABC is similar both to triangle ABD and to triangle ACD (because corresponding angles are equal), so their corresponding sides are proportional. In particular, $a/c = c/y$ and $a/b = b/(a - y)$. That is, $ay = c^2$ and $a(a - y) = a^2 - ay = b^2$. So $a^2 - c^2 = b^2$.

The proof above is based on the same diagram used to prove Proposition VI,31. But neither the statement of the Pythagorean Theorem in **3.** (the form in which it is usually stated), nor the proof just given, makes any reference to *areas*; only *length* relationships are involved. Thus both the *meaning* of the proposition stated in **3.** and the argument used to justify it are conceptually distinct from Euclid's propositions I,47 and VI,31. Nonetheless, it is worth noting in passing the connection between Figure 5.3 and the bottom part of Euclid's 'windmill' diagram (Figure 5.1). For if we modify Figure 5.3 by erecting the square $BCFE$ on $\overline{BC}$ and extending $\overline{AD}$ to meet the opposite side of that square at G (Figure 5.5), then the rectangle $BDGE$ has height a and width y, and the first proportion used in the proof of **3.** shows that $y = c^2/a$; so $BDGE$ has area c^2. Likewise, rectangle $DCFG$ has area b^2, since it has height a and width $a - y = b^2/a$.

The proof given for **3.** is attractive from a pedagogical standpoint: It is concise, the diagram is much simpler than that for Euclid's proof of I,47, and the argument is easy for students to follow (much more so than that for VI,31, because both the idea of a special case implying the truth of a more general statement and the recognition that the triangles in Figure 5.3 *are* an instance of the configuration described in the statement of VI,31 are difficult for novices to grasp). However, even after being

Fig. 5.5 The base of the 'windmill' diagram

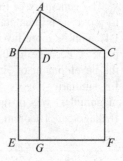

Fig. 5.6 $\cos^2 \theta + \sin^2 \theta = 1$

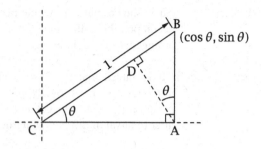

given Figure 5.3 and told to invoke properties of similar triangles, students may have difficulty discovering that proof, since they may have trouble finding the *right* set of proportional relations to use (a difficulty that may nonetheless make the exercise a valuable one for improving students' appreciation of the effort and creativity involved in finding proofs).

4. The trigonometric form of the Pythagorean Theorem: Let θ be an acute angle in a right triangle. Then $\cos^2 \theta + \sin^2 \theta = 1$.

Proof: Since ratios of sides are unaffected by scaling, it suffices to consider a right triangle ABC with hypotenuse of length 1 and right angle at A. Let angle θ be at vertex C, place it (for convenience) in standard position on a rectangular coordinate system, and erect the altitude $|\overline{AD}|$ (Figure 5.6). In triangle ACD, $\cos \theta = |\overline{CD}|/|\overline{AC}| = |\overline{CD}|/\cos \theta$, and in triangle BAD, $\sin \theta = |\overline{BD}|/|\overline{AB}| = |\overline{BD}|/\sin \theta$, since angle $BAD = \theta$. So $1 = |\overline{CD}| + |\overline{BD}| = \cos^2 \theta + \sin^2 \theta$.

Apart from its orientation and scaling, Figure 5.6 is the same as Figure 5.3, and if the labels a, b, c, and y are defined as in **3.**, then $\cos \theta = b/a$ and $\sin \theta = c/a$, whence $ay = c^2$ and $a(a - y) = b^2$, as in the earlier proof. Should the proof of **4.** then be regarded merely as a variant of the proof of **3.**?

The situation is similar to the relation between the gnomon and induction proofs considered in Chapter 3. The trigonometric proof of statement **4.** is computationally simpler than the algebraic proof of **3.**, even though algebraically equivalent to it; and the geometric representation of the expression $\cos^2 \theta + \sin^2 \theta$ as the length of the hypotenuse adds perspicuity to the proof of **4.** as well. Conceptually, then, the two arguments are distinct: an example of how a judicious choice of primitive concepts (here, ratios of lengths rather than lengths themselves) can make a proof both easier to carry out and easier to understand and remember.

5.1 Two dissection proofs

Dissection proofs, used to show the equality of areas, are of two kinds:

Either   (i) It is shown that two geometric figures of *different* shapes can be decomposed into the *same* set of non-overlapping pieces, differently arranged. (Such figures are said to be *equidecomposable*.)

or   (ii) It is shown that the *same* figure can be decomposed into two *different* sets of non-overlapping pieces, each set containing some pieces congruent to pieces in the other, so that equal areas are left after removal of different congruent pieces.

Such proofs have surprisingly wide applicability, in view of

The Bolyai-Gerwin Theorem: Any two polygons of equal area are equidecomposable.[6]

The two proofs that follow are of the second type.

5. A proof without words (See Figure 5.7.)

This ancient dissection proof compels immediate assent, even from young students with no algebraic background, that the area of the square on the hypotenuse of a right triangle is equal to the sum of the areas of the squares on the other two sides. To understand it, all that is required is knowledge that the angles opposite the legs in a right triangle are complementary (in order to confirm that the sides of the large square in the left diagram are straight line segments).

The diagram on the right is in effect a geometric representation of the identity $(a+b)^2 = a^2 + b^2 + 2ab$, so given the latter, only the diagram on the left is needed to carry out the proof. On the other hand, with a slight alteration, the two diagrams in Figure 5.7 may be adapted to serve a quite different purpose: that of showing that $\sin(\alpha + \beta) = \sin\alpha\cos\beta + \sin\beta\cos\alpha$ (See Figure 5.8 below, taken from

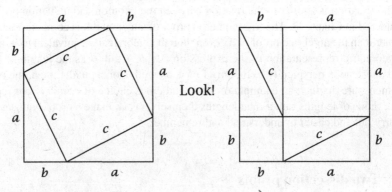

Fig. 5.7 A simple dissection proof

---

[6]For a proof, see Boltyanskii (1963).

$$\sin{(\alpha + \beta)} = \sin{\alpha}\cos{\beta} + \sin{\beta}\cos{\alpha}$$

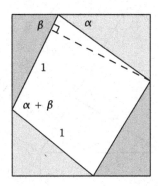
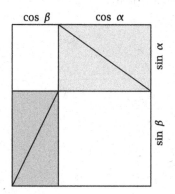

Fig. 5.8 Sine of the sum of two angles

Nelsen 2000, p. 40, where the figure is credited to Volker Priebe and Edgar A. Ramos. ©The Mathematical Association of America 2013. All rights reserved.)

A variation of dissection proof **5.**, due to the twelfth-century Indian mathematician Bhāskara, is also well-known, and a diagram much like that on the right side of Figure 5.7 is used to illustrate proposition II,4 of Euclid's *Elements*. So the question again arises, could not Euclid have proved proposition I,47 more simply by using those diagrams? Heath thought that the only objection to that idea was that such dissection proofs had "no specifically Greek" character (Heath 1956, vol. I, p. 355). Knorr, however, considered that objection "unjust" (Knorr 1975, p. 178 and fn 18 thereto, pp. 204–5). The idea that the passage from the left to the right diagram in Figure 5.7 requires spatial translation of the constituent triangles, an operation not justified by Euclid's axioms, also does not hold up to scrutiny, for two reasons. First, Euclid's proof of I,47 relies on proposition I,4 (the side-angle-side criterion for congruence), whose proof, as noted earlier, itself involves spatial displacement of a figure. Second, the two diagrams in Figure 5.7 can be superimposed in a single diagram that is constructible by Euclidean methods from a given right triangle ABC. (See the remark following the next proof.)

6. A proof involving congruent pentagons: Given right triangle ABC with right angle at C, construct squares on sides $\overline{BC}, \overline{AC}$, and $\overline{AB}$ and label them 2, 3, and 4, respectively. (See Figure 5.9.) Extend the side of square 2 opposite $\overline{BC}$ and the side of square 3 opposite $\overline{AC}$ until they meet, and draw the diagonal from C to that intersection point. Label the resulting triangles 5 and 6. Label the vertices of square 4 diagonally opposite to A and B as D and E, respectively, and draw the perpendicular from D to the extension of $\overline{CB}$ and the perpendicular from E to the extension of $\overline{CA}$. Label the resulting triangles on $\overline{AE}$ and $\overline{BD}$ as 7 and 8, respectively. Regions 1, 2, 3, 5, and 6 together form a pentagon that is *congruent* to the pentagon formed by regions 1, 4, 7, and 8; and since triangles 5, 6, 7, and 8 are each congruent to triangle ABC, if triangles 1, 5, and 6 are removed from the first

Fig. 5.9 Overlapping
congruent pentagons

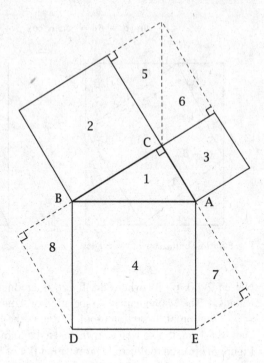

of those pentagons and triangles 1, 7, and 8 from the second, the remaining areas
must be equal. That is, the area of square 4 equals the sum of the areas of squares
2 and 3.   q.e.d.

Note that in both proofs **5.** and **6.** triangles congruent to the original are constructed
on the sides of the squares in question. But there are only five triangles in Figure 5.9,
while there are eight in the two diagrams of Figure 5.7.

As noted earlier, it is possible to combine the two diagrams of Figure 5.7 into
one figure and then carry out the dissection. In particular, if two further triangles
congruent to ABC are added to Figure 5.9, one (9) below $\overline{AB}$ oriented so as to form
a rectangle with ABC and the other (10) in the same orientation below $\overline{DE}$, the
resulting Figure 5.10 may be viewed as the two squares of Figure 5.7 placed so as
to overlap in triangle ABC.

The last of the proofs to be considered here is another 'proof without words'
(the forty-first of those on Bogomolny's site, credited there to Geoffrey Margrave
of Lucent Technologies) which, like proof **5.**, requires only knowing that the angles
opposite the legs in a right triangle are complementary. Unlike the dissection proof,
however, it is based on *scaling* (an operation performable by Euclidean means
according to proposition VI,12 of the *Elements*). Like the 'folding bag' proof
in Figure 5.4, three triangles each similar to the original are employed, but the
conclusion results not from overlap of their areas, but from the equality of the
lengths of opposite sides of a rectangle.

Fig. 5.10 The two parts of
Figure 5.7 combined

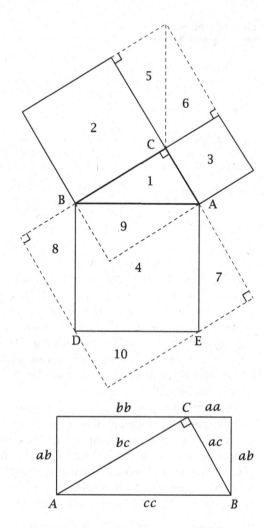

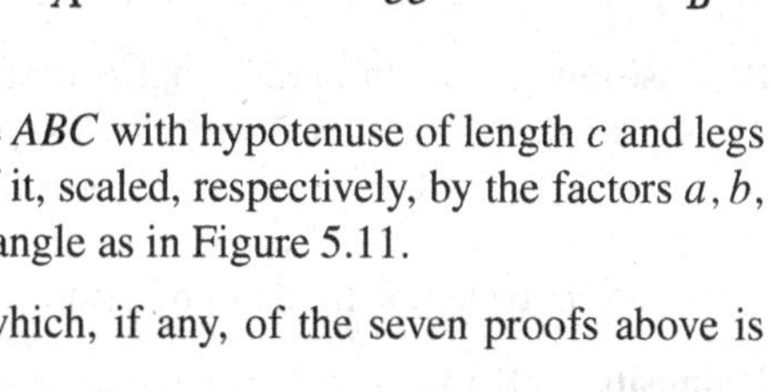

Fig. 5.11 Three scaled
copies fitted together

7. Proof by scaling: Given right triangle *ABC* with hypotenuse of length *c* and legs
of lengths *a* and *b*, make three copies of it, scaled, respectively, by the factors *a*, *b*,
and *c*, and assemble them to form a rectangle as in Figure 5.11.

Readers may judge for themselves which, if any, of the seven proofs above is
simplest or most perspicuous.

5.2 Further consequences and extensions

Euclid's Proposition VI,31 extended Proposition I,47 to 'arbitrary' similar figures
described on the sides of a right triangle. In another direction, the Law of Cosines
provides an extension of I,47 to arbitrary triangles. It includes the Pythagorean
Theorem as a special case and also implies its converse.

Fig. 5.12 The converse of
the Pythagorean Theorem

Fig. 5.13 The Law of
Cosines derived from the
Pythagorean Theorem

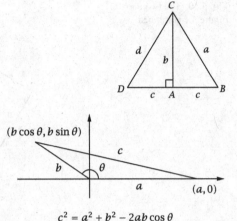

$$c^2 = a^2 + b^2 - 2ab\cos\theta$$

In Euclid's treatment, the converse is established in Proposition I,48 (the last in Book I) as an easy consequence of I,47. Specifically (see Figure 5.12), given a triangle ABC in which the square on one of the sides (say $\overline{BC}$) is equal to the sum of the squares on the other two sides, construct a segment perpendicular to $\overline{AC}$ at A of length $c = |\overline{AB}|$ and join its terminal point D to C. Letting $a = |\overline{BC}|, b = |\overline{AC}|$, and $d = |\overline{CD}|$, then $a^2 = b^2 + c^2$ by assumption, and $b^2 + c^2 = d^2$ by I,47. So $a = d$ and triangle ACD is congruent to triangle ABC by the side-side-side criterion (I,8). Hence angle BAC is right.

The Law of Cosines is also a consequence of the Pythagorean Theorem. Today, after extending the definition of the trigonometric functions to the interval $[0, \pi/2]$, the proof is usually carried out by applying the distance formula to a triangle positioned as in Figure 5.13 (which illustrates the obtuse-angled case). The Law of Cosines in its trigonometric form of course does not appear in the *Elements*. But geometric equivalents for the obtuse- and acute-angled cases are stated and proved as propositions 12 and 13 in Book II. Their statements are:

Proposition II, 12: In obtuse-angled triangles the square on the side subtending the obtuse angle is greater than the [sum of the] squares on the sides containing the obtuse angle by twice the rectangle containing one of the sides, namely that on [the extension of which] the perpendicular falls, and the straight line cut off [from that extension] outside [the triangle] by [that] perpendicular. (Figure 5.14(a))

Proposition II, 13: In triangles [containing an acute angle], the square on the side subtending [that] acute angle is less than the [sum of the] squares on the sides containing [that] angle by twice the rectangle contained by one of the sides about [that] acute angle, namely that on which the perpendicular falls, and the straight line cut off [on that side] within [the triangle] by the perpendicular towards [that] acute angle. (Figure 5.14(b))

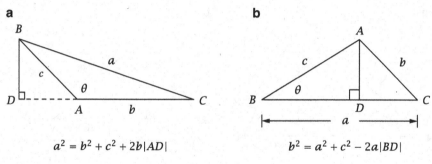

$$a^2 = b^2 + c^2 + 2b|AD| \qquad\qquad b^2 = a^2 + c^2 - 2a|BD|$$

Fig. 5.14 Euclid's propositions II,12 and II,13

Euclid's proofs of those propositions both rely on the Pythagorean Theorem, together with geometric analogs (Propositions II,4 and II,7) of the algebraic identities $(a + b)^2 = a^2 + b^2 + 2ab$ and $(a - b)^2 = a^2 + b^2 - 2ab$. However, in his commentaries on Propositions II,12 and II,13, Heath shows that each of those results can alternatively be proved in the same manner as I,47, using variants of the 'windmill' diagram (Heath 1956, vol. I, pp. 404–405 and 407–408. For Proposition II,13 there are three cases to consider, depending on whether one of the other angles is right, one is obtuse, or all are acute.) Thus, for triangles not containing a right angle, the Law of Cosines can be proved independently of, but by the same method as, Euclid's first proof of the Pythagorean Theorem.

Very recently, a uniform proof of the Law of Cosines, valid for all angles and independent of the Pythagorean Theorem, has been given by John Molokach (http://www.cut-the-knot.org/pythagoras/CosLawMolokach.shtml). It is presented here as a final example of a simple, conceptually distinct proof of a statement that implies the Pythagorean Theorem.

8. Direct derivation of the Law of Cosines: Given any triangle ABC, let a, b, and c denote the sides opposite angles A, B, and C, respectively. At least two of the angles, say A and B, must be acute. Then (see Figure 5.15), regardless of whether angle C is acute, right, or obtuse:

(1) $$a = b \cos C + c \cos B,$$

(2) $$b = a \cos C + c \cos A,$$

(3) $$c = a \cos B + b \cos A.$$

Multiplying (1) by a, (2) by b and (3) by c then gives

(4) $$a^2 = ab \cos C + ac \cos B,$$

(5) $$b^2 = ab \cos C + bc \cos A,$$

(6) $$c^2 = ac \cos B + bc \cos A.$$

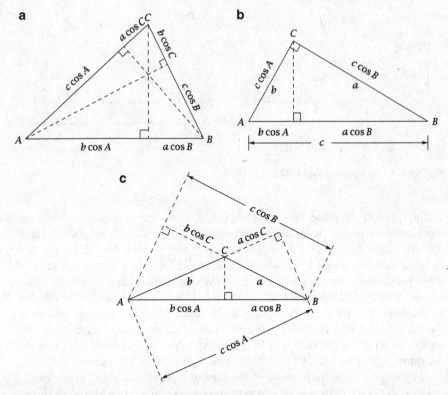

Fig. 5.15 (a) The Law of Cosines: all angles acute. (b) The Law of Cosines: angle C right. (c) The Law of Cosines: angle C obtuse

Subtracting any two of these equations from the third then gives one of the three forms of the Law of Cosines; e.g., subtracting (4) and (5) from (6) yields that $c^2 - a^2 - b^2 = -2ab\cos C$.   q.e.d.

References

Bogomolny, A.: The Pythagorean Theorem and its many proofs. http://www.cut-the-knot.org/pythagoras/index.shtml (2012)

Boltyanskii, V.G.: Equivalent and Equidecomposable Figures. D.C. Heath, Boston (1963)

Heath, T.: The Thirteen Books of Euclid's Elements (3 vols.). Dover, New York (1956)

Knorr, W.: The Evolution of the Euclidean Elements. A Study of the Theory of Incommensurable Magnitudes and its Significance for Early Greek Geometry. Reidel, Dordrecht (1975)

Loomis, E.S.: The Pythagorean Proposition—its Demonstrations Analyzed and Classified, and Bibliography of Sources for Data of the Four Kinds of Proofs, 2nd ed. NCTM, Washington, D.C. (1940)

Maor, E.: The Pythagorean Theorem, a 4000-year History. Princeton U.P., Princeton (2007)

Nelsen, R.B.: Proofs without Words II. More Exercises in Visual Thinking. Math. Assn. Amer., Washington, D.C. (2000)

Yanney, B.F., Calderhead, J.A.: New and old proofs of the Pythagorean Theorem. Amer. Math. Monthly **3**, 65–67, 110–113, 169–171, 299–300; **4**, 11–12, 79–81, 168–170, 250–251, 267–269; **5**, 73–74; **6**, 33–34, 69–71 (1896–9)

Zimba, J.: On the possibility of trigonometric proofs of the Pythagorean Theorem. Forum Geometricorum **9**, 275–278 (2009)

Chapter 6
The Fundamental Theorem of Arithmetic

The Fundamental Theorem of Arithmetic (FTA) states that every integer greater than 1 has a factorization into primes that is unique up to the order of the factors. The theorem is often credited to Euclid, but was apparently first stated in that generality by Gauss.[1] Note that the statement has two parts: First, every integer greater than 1 *has* a factorization into primes; second, any two factorizations of an integer greater than 1 into primes must be identical except for the order of the factors. The proofs of each of those parts will thus be considered separately.

The first part is the subject of propositions VII,31 and VII,32 of the *Elements*. Proposition VII,31 states that any composite number (that is, any number that has a proper divisor other than one) is divisible by some prime. Having established that, Euclid then immediately concludes in proposition VII,32 that any number greater than 1 is either prime or is divisible by some prime.

Euclid's proof of VII,31: Let A be a composite number. By definition, A has a proper divisor B other than one. If B is prime, we are done. If not, B has a proper divisor C other than one, and then C is a proper divisor of A. If C is prime, we are done. Otherwise, C has a proper divisor other than one. Continuing in this fashion, one must eventually obtain a prime divisor of A, since otherwise there would be an infinite sequence of divisors $B, C, \ldots$ of A, each smaller than the one before, which is impossible.

Second proof (of VII,31 and VII,32 together): By complete induction on the integer $A > 1$. Suppose every integer greater than 1 and less than A is divisible by some prime. Consider A. If A is prime, we are done. Otherwise, $A = BC$ with $1 < B, C < A$. By the inductive hypothesis, B is divisible by some prime, and that prime divides A.

---

[1] In the *Disquisitiones Arithmeticae* (Gauss 1801, Bd. I, p. 15). (See Collison 1980, p. 98.) However, the result was certainly known, if not explicitly stated, beforehand. For example, Euler used it implicitly in his 1737 proof of the infinitude of the primes (via the divergence of the harmonic series).

© Springer International Publishing Switzerland 2015
J.W. Dawson, Jr., *Why Prove it Again?*, DOI 10.1007/978-3-319-17368-9_6

Repeated application of VII,32 then establishes the existence of a prime factorization for any integer greater than 1.

Note that the first of the proofs above is a *reductio*, while the second is a direct
proof that explicitly uses induction. Euclid takes for granted that there cannot be an
infinite strictly decreasing sequence of positive integers — a statement that nowadays would be deemed to require proof. The most direct proof is by means of the
well-ordering principle, which is equivalent to induction (or complete induction).
Indeed, the statement in question is itself logically equivalent to induction.

There is little more development of the first part of the FTA, since the second
argument above is so simple. (Some further remarks about the first part will,
however, be made at the end of the chapter). Consider then the second part of
the FTA.

A corollary of the FTA is Euclid's Lemma, which asserts that if a prime divides
a product it must divide one of the factors. A strong form of Euclid's Lemma, but
restricted to products of just two factors, was stated by Euclid as proposition VII,30
of the *Elements*. A consequence of that result is Euclid's proposition IX,14 ('If a
number be the least that is measured by [three distinct] prime numbers, it will not
be measured by any other prime number except those originally measuring it.').

Euclid did not consider products of more than three primes, nor products
involving repeated factors. However, his proof of IX,14 can be applied to conclude
that the representation of a number as a product of *distinct* primes is unique except
for the order of the factors. To extend to products involving repeated factors one
can apply proposition IX,13 of the *Elements* (which states that the only divisors
of p^k are the numbers $1, p, p^2, \ldots, p^k$) in combination with VII,30. Alternatively,
one can argue, as Gauss later did, that if a prime p appears to the power j in one
factorization of a number n and to the power k in another, with $j \leq k$, then dividing
by p^j will yield two factorizations of another number, at most one of which involves
the factor p. Applying VII,30 repeatedly then shows that p must in fact occur in
neither, so $j = k$. Any other repeated prime factors may be similarly eliminated, so
only products of distinct prime factors need be considered.

Consequently, Euclid's Lemma also implies the second part of the FTA, and
it is useful to distinguish proofs of the second part of the FTA that do *not* first
prove Euclid's Lemma from those that do. Proofs of the second part of the FTA
may also be distinguished according to what extent (if any) they use mathematical
induction, whether they are direct or indirect, whether they invoke the concepts of
least common multiple or greatest common divisor, and whether they employ the
division algorithm, the Euclidean algorithm, or neither.

6.1   Direct proofs of Euclid's Lemma

The statement of proposition VII,30 in Euclid's *Elements* is just that of Euclid's
Lemma: 'If two numbers by multiplying one another make some number, and
any prime number measure the product, it will also measure one of the original

numbers.' The *proof*, however, only uses the assumption that the number measuring the product is prime to deduce that it is *relatively* prime to each of the original factors. It thus establishes the following stronger result, stated earlier in Chapter 4.

Theorem: *If a divides bc and is relatively prime to b, then a divides c.*

Proof (a modern paraphrase of Euclid's argument): Suppose a divides bc, say $bc = ad$, and a is relatively prime to b. Then a must be the least natural number that, when multiplied by d, yields a multiple of c; that is, ad must be the least common multiple of c and d. For let f denote the least such number, and suppose $fd = ec$. By the division algorithm, $a = qf + r$ for some q, r with $0 \leq r < f$, so $bc = ad = qfd + rd = qec + rd$. Hence $rd = bc - qec = (b - qe)c$. By the minimality of f, r must equal 0, so $a = qf$ and (since $c \neq 0$) $b = qe$. q is thus a common factor of a and b, which implies that $q = 1$. Therefore $a = f$, as claimed.

To finish the proof, Euclid appealed to his proposition VII,20 (whose proof, however, was faulty; cf. footnote 1 in Chapter 4): a is the least natural number for which $a/b = c/d$, so a divides c. Alternatively, one may apply the division algorithm again to deduce that $c = pa + s$ for some p, s with $0 \leq s < a$. Then, $s = c - pa$, so

$$sd = cd - pad = cd - pbc = (d - pb)c.$$

That is, sd is a multiple of c, so by the minimality of a, $s = 0$. Thus $c = pa$, so a divides c.   q.e.d.

Whichever method is used to complete the proof above, the argument as a whole invokes the division algorithm twice, since (again as noted in footnote 1 of Chapter 4) a correct proof of Euclid's VII,20 employs the division algorithm.

By contrast, the next proof (from Rademacher and Toeplitz 1957, pp. 71–72) of the weaker form of Euclid's Lemma does so only once, to show that the least common multiple of two numbers divides any common multiple of them, but it makes use of an additional fact (displayed as (7) below) not employed in Euclid's argument.

Second proof: If m is the least common multiple of two numbers a and b and M is any common multiple of them, then m divides M; for, by the division algorithm, $M = qm + r$ for some q and r with $0 \leq r < m$, so the minimality of m implies that $r = M - qm$ must be 0. In particular, m must divide ab, say $md = ab$, where

(7) d must divide both a and b.

(For if $m = ka = lb$, then $md = kad = ab$ and $md = lbd = ab$, so $kd = b$ and $ld = a$.) Now suppose the prime p divides BC, and let L be the least common multiple of p and B. Since BC and pB are both common multiples of p and B, L divides both of those products — say $LE = BC$ and $LF = pB$. By (7) above, F divides both p and B, and since p is prime, either $F = 1$ or $F = p$; that is, either $L = pB$ or $L = B$. In the former case, $LE = pBE = BC$, so $pE = C$, that is, p divides C. In the latter case, p divides B, since L is a multiple of p.   q.e.d.

That the quantity d in (7) is actually the *greatest* common divisor of a and b is nowhere used in the proof above. However, Euclid's Lemma is an almost immediate consequence of the following well-known characterization of greatest common divisors.

Linear representation theorem: *If d is the greatest common divisor of a and b, then there are integers m and n, exactly one of which is positive, for which $d = ma + nb$.*

Proof of Euclid's lemma from the linear representation theorem: If p is a prime that divides bc but not b, then the greatest common divisor of p and b is 1. Therefore $1 = mp + nb$ for some integers m and n, so $c = mpc + nbc$. Since p divides each summand on the right, p divides c. (Exactly the same argument holds if p is not necessarily prime, but merely relatively prime to b.)

The argument just given forms the conclusion of two distinct proofs of Euclid's Lemma, which differ in how the linear representation theorem itself is derived.

Third proof (summarized from Courant and Robbins 1941, pp. 45–47): The representation of the greatest common divisor of integers a and b as an integral linear combination of them is obtained constructively by examining the proof of the Euclidean algorithm (proposition VII,2 in Euclid's *Elements*). That proof, and the implementation of the algorithm to compute m and n explicitly, involves iterated application of the division algorithm, in which *the number of iterations required is not fixed*, as in the two proofs given earlier, but depends on the values of a and b.[2]   q.e.d.

Alternatively, the linear representation of the greatest common divisor may be demonstrated non-constructively as follows.

Fourth proof: The set I of *all* linear combinations $ma + nb$, as m and n range over all integers, is an ideal within the ring of integers. Let d be an element of I whose absolute value is minimal. A single application of the division algorithm shows that d must divide every element of I, so in particular it must divide a and b. But *any* common divisor of a and b must also divide every element of I, including d. Therefore d must be a greatest common divisor of a and b (as must $-d$, so d may be taken to be positive without loss of generality).   q.e.d.

---

[2]Of course, the division algorithm itself involves iterated *subtraction*, where the number of iterations likewise depends on the values of the dividend and divisor.

A priori, the greatest common divisor of a and b is just that, the common divisor which is the largest. But one important consequence of the linear representation theorem is the following property of the greatest common divisor.

Divisibility property of the gcd: The greatest common divisor of a and b is divisible by every common divisor of a and b.

Proof. Let c be a common divisor of a and b. Express the greatest common divisor d of a and b as $d = ma + nb$. Then, since c divides both a and b, c divides d as well.

Weintraub (in Weintraub 2008) gave the following proof of Euclid's lemma from the divisibility property of the gcd (which itself may be proved in various ways).

Fifth proof: Consider ac and bc. They have a greatest common divisor d. Now c divides both ac and bc, so by the divisibility property of the gcd, c divides d. Write $d = cz$. Now d divides bc, that is, cz divides bc, so z divides b. Similarly, cz divides ac, so z divides a. But a and b are assumed to be relatively prime, so $z = 1$ and $d = c$. Now a certainly divides ac, and a divides bc by hypothesis, so a divides d by the divisibility property of the gcd again; since $c = d$, a divides c.

6.2   Indirect proofs of the FTA and Euclid's Lemma

The Fundamental Theorem of Arithmetic may also be proved outright, without first proving Euclid's Lemma, through inductive arguments by *reductio*. Two such proofs, the first by Ernst Zermelo and the second by Gerhard Klappauf,[3] are reproduced in Scholz (1961). Both begin by presuming, contrary to the statement of the FTA, that there are integers with distinct prime factorizations, among which there must be some least integer m. Zermelo then argued as follows.

Sixth proof: Suppose $m = p_1 p_2 \cdots p_k = q_1 q_2 \cdots q_s$, with $p_1 \leq p_2 \leq \cdots \leq p_k$ and $q_1 \leq q_2 \leq \cdots \leq q_s$. By the minimality of m, $p_1 \neq q_1$, so without loss of generality we may suppose that $p_1 < q_1$. Then the number

$$n = m - p_1 q_2 \cdots q_s = p_1 (p_2 \cdots p_k - q_2 \cdots q_s) = (q_1 - p_1)(q_2 \cdots q_s)$$

is less than m, and so must possess a unique prime factorization. Since p_1 is less than every q_i, it must therefore divide $q_1 - p_1$. But then p_1 would divide q_1, which is prime. Since $1 < p_1 < q_1$, that is impossible.   q.e.d.

In the paper in which he presented the proof just given, Zermelo stated that his reason for doing so was to show that even in elementary number theory it was

---

[3]Published originally in Zermelo (1934) and Klappauf (1935), respectively.

possible to simplify the proofs.[4] His proof, in turn, then stimulated Klappauf to show that the method Zermelo had used to produce the counterexample n could be further simplified.

Seventh proof: Let m be as in Zermelo's proof and consider the remainders r_i, for $i = 1, \ldots, s$ that are obtained when each q_i is divided by p_1. We have $q_i = a_i p_1 + r_i$, where each $r_i < p_1$. Since $p_1 < q_i$ for each i, every a_i must be positive; and since each q_i is a prime different from p_1, every r_i is also positive. Hence $m = q_1 q_2 \cdots q_s$ can be written as $m = A p_1 + R$, where $R = r_1 r_2 \cdots r_s$ and A and R are both positive. Since p_1 divides m, it must also divide R. But p_1 cannot divide any r_i; so factoring each r_i into primes yields a factorization of R that is distinct from the factorization involving p_1. Since $R < m$ that contradicts the minimality of m.   q.e.d.

Unlike Zermelo's proof, Klappauf's employs the division algorithm. Moreover, in Klappauf's proof $r_1 = q_1 - a_1 p_1 \leq q_1 - p_1$, and $r_i < q_i$ for $i \geq 2$, so the number R used therein to contradict the minimality of m is less than the number $n = (q_1 - p_1) q_2 \cdots q_s$ used for that purpose in Zermelo's proof.

Euclid's Lemma may also be proved by *reductio*. Indeed, Gauss did so (for the contrapositive statement) in his *Disquisitiones Arithmeticae*. His proof, presented next below, is actually a double *reductio* that invokes the division algorithm thrice.

Eighth proof: Gauss first showed by *reductio* that no prime p can divide a product of two smaller positive integers. For suppose to the contrary that p is a prime that divides such a product, and let $r < p$ be the least positive integer for which there exists a positive integer $s < p$ such that p divides rs. Then $r \neq 1$ (since $s < p$), so r does not divide the prime p. Hence by the division algorithm, $p = qr + t$, where $0 < t < r$. But then $ts = ps - qrs$ is divisible by p, contrary to the minimality of r.

To complete the proof of Euclid's Lemma, suppose then (again by *reductio*) that a prime p divides bc but neither b nor c. Then the division algorithm gives $b = q_1 p + r_1$ and $c = q_2 p + r_2$, with $0 < r_1, r_2 < p$. So bc can be expressed in the form $qp + r_1 r_2$. That is, $r_1 r_2 = qp - bc$, which is a multiple of p if bc is; but that contradicts Gauss's earlier result.   q.e.d.

Another *reductio* proof of Euclid's Lemma, in the strong form stated by Euclid, was given by Daniel Davis and Oved Shisha in a little-known article in *Mathematics Magazine* (Davis and Shisha 1981).[5] The last of the proofs to be considered here, it is an elegant exemplar of purity of method.

---

[4]He noted that he had first communicated his proof around 1912, in correspondence with A. Hurwitz, E. Landau and others, and was stimulated to publish it after reading the proof in the German edition (1933) of Rademacher and Toeplitz's book (the second proof given above), unaware when he did so that a proof similar to his had been published six years earlier by Helmut Hasse (Hasse 1928). In addition, F.A. Lindemann had published another similar, but somewhat more complicated, proof just the year before (Lindemann 1933).

[5]Their paper actually gave five slightly variant proofs.

Ninth proof: Assume that there is a triple (A, B, C) of positive integers for which both of the following properties hold:

$P_1(A, B, C)$. A divides BC but is relatively prime to B.
$P_2(A, B, C)$. A divides BC but does not divide C.

Then for $i = 1, 2$, it follows directly that

$(1.i)$ $\qquad$ If $P_i(A, B, C)$ and $B > A$, then $P_i(A, B - A, C)$

and

$(2.i)$ $\qquad$ If $P_i(A, B, C)$, say $BC = AD$, then $P_i(B, A, D)$.

Proof of (1.1): If A divides BC and $B > A$, then A divides $(B-A)C = BC-AC$; and if A is relatively prime to $B = (B - A) + A$ it must also be relatively prime to $B - A$.

Proof of (1.2): If A divides BC but not C, then A divides $(B - A)C = BC-AC$ but not C.

Proof of (2.1): If $BC = AD$ and A is relatively prime to B, then B divides AD and is relatively prime to A.

Proof of (2.2): If $BC = AD$ but A does not divide C, then B divides AD but does not divide D.

Among all triples (A, B, C) satisfying P_1 and P_2 there is at least one, say (A_1, B_1, C_1), that minimizes $A + B + C$. Then by P_2, $A_1 \neq 1$, so by P_1, $A_1 \neq B_1$. By $(2.i)$, the triple (A_1, B_1, D) also satisfies P_1 and P_2, and $B_1 C_1 = A_1 D$; so if $B_1 < A_1$, then $D < C_1$ and therefore $A_1 + B_1 + D < A_1 + B_1 + C_1$, contrary to the minimality property of (A_1, B_1, C_1). The only remaining possibility is $B_1 > A_1$. Then by $(1.i)$, the triple (A_1, B_1-A_1, C_1) also satisfies P_1 and P_2; but $B_1-A_1 < B_1$, so $A_1+(B_1-A_1)+C_1 = B_1+C_1 < A_1+B_1+C_1$, again contrary to the minimality property of (A_1, B_1, C_1). Hence by *reductio*, no triple (A, B, C) satisfying both P_1 and P_2 exist. That is, if A divides BC but is relatively prime to B, then A must divide C. $\qquad$ q.e.d.

This proof of Davis and Shisha is distinguished above all by its economy of means, for it employs nothing more than subtraction and the concepts involved in the statement of Euclid's Lemma (divisibility and relative primality).

6.3 Summary

It should be clear from the commentary above that the two proofs of the first part of the FTA considered in this chapter, and the nine proofs of the second part, are all structurally distinct. Moreover, they exemplify several of the rationales for presenting alternative proofs enumerated in Chapter 2: the desires to simplify, to minimize conceptual prerequisites, to extend to broader contexts, to achieve methodological purity, and to find new routes to a goal. As to proofs of the first part, the second proof is direct while the first proof is a proof by *reductio*. As to the second part, Gauss's proof extended that of Euclid; the second, sixth and seventh proofs, and especially the ninth, exhibit various forms of simplification; and the third and fourth proofs introduce a concept (that of representing the greatest common divisor of two integers as a linear combination of them) that is foreign to all the others, while the fifth proof deliberately avoids using that concept.

It is enlightening to examine these proofs in the context of generalizations to commutative ring theory. As to the proofs of the first part, the first proof leads directly to the concept of a Noetherian ring, and directly generalizes to show that every element in a Noetherian integral domain has a (that is, at least one) factorization into primes. The second proof, while simpler, is one that is restricted to the positive integers.

As to the proofs of the second part, again the most direct proofs, the sixth, seventh, and ninth, are restricted to the positive integers.

The other proofs generalize to commutative rings of various kinds. By definition, a Euclidean domain is one in which there is a division algorithm, properly interpreted, and these proofs show that Euclid's lemma holds in Euclidean domains, and hence that these are unique factorization domains (that is, that the analog of the FTA holds in them). By definition, a principal ideal domain is one in which an appropriate generalization of the linear representation theorem holds, and so the third and fourth proofs, which rely on that concept, show that every principal ideal domain is a unique factorization domain. Since there are principal ideal domains that are not Euclidean, this provides a further generalization. Furthermore, not every unique factorization domain is a principal ideal domain, so the fifth proof generalizes still further (though in this case one needs some other argument to show that the divisibility property of the gcd holds). The second proof, which introduces the concept of the least common multiple, is stated for the positive integers, and directly generalizes to Euclidean domains. But that same concept is fruitful in the more general contexts we have just described, and that proof can be modified to be valid in them as well.

References

Collison, M.J.: The unique factorization theorem, from Euclid to Gauss. Math. Mag. **53**(2), 96–100 (1980)

Courant, R., Robbins, H.: What is Mathematics? Oxford U., Oxford et al. (1941)

Davis, D., Shisha, O.: Simple proofs of the fundamental theorem of arithmetic. Math. Mag. **54**, 18 (1981)

Gauss, C.F.: Disquisitiones Arithmeticae (1801). In his Werke, I, 1–466. W.F. Kaestner, Göttingen (1863)

Hasse, H.: Über eindeutige Zerlegung in Primelemente oder in Primhauptideale in Integritäts Bereichung. J. reine. u. angew. Math. **159**, 3–12 (1928)

Klappauf, G.: Beweis des Fundamentalsatz der Zahlentheorie. Jahresb. DMV **45**, 130 (1935)

Lindemann, F.A.: The unique factorization of a positive integer. Quarterly J. Math. **4**, 319–320 (1933)

Rademacher, H., Toeplitz, O.: The Enjoyment of Mathematics. Princeton U.P., Princeton (1957)

Scholz, A.: Einführung in die Zahlentheorie. De Gruyter, Berlin (1961)

Weintraub, S.H.: Factorization: Unique and Otherwise. AK Peters, New York (2008)

Zermelo, E.: Elementare Betrachtungen zur Theorie der Primzahlen. Nachr. Gesell. Wissen. Göttingen (Neue Folge) **1**, 43–46 (1934)

Chapter 7
The Infinitude of the Primes

Euclid's proof that the prime numbers are "more than any assigned multitude" (*Elements*, proposition IX, 20) has long been hailed as a model of elegance and simplicity. Yet, surprisingly, it has also been misrepresented in a great many accounts: The article Hardy and Woodgold (2009) gives a detailed list of sources, including many by eminent number theorists, that either erroneously describe the structure of Euclid's proof or make false historical claims about it. It is wise, therefore, to begin by quoting Euclid's argument directly, as it is given in Heath's translation (Heath 1956, vol. II, p. 412).

Proposition IX, 20: Prime numbers are more than any assigned multitude of prime numbers.

Proof: Let A, B, C be the assigned prime numbers. I say that there are more prime numbers than A, B, C. For let the least number measured by them be taken and let it be [represented by the line segment] DE; [then] let the unit [segment] DF be added to DE. Then EF is either prime or not. First, let it be prime. Then A, B, C, EF have been found which are more than A, B, C. Next, let EF not be prime; therefore [by proposition VII,31] it is measured by some prime number. Let it be measured by the prime number G. I say that G is not the same as any of the numbers A, B, C. For, if possible, let it be so. Now A, B, C measure DE; therefore, G will also measure DE. But it also measures EF. Therefore G, being a number, will measure the remainder, the unit DF, which is absurd. Therefore G is not the same as any of the numbers A, B, C. And by hypothesis it is prime. Therefore the prime numbers A, B, C, G have been found, which are more than the assigned multitude of A, B, C.

Commentary: It is to be noted, first of all, that Euclid speaks of exactly three 'assigned' prime numbers. But it is clear from the statement of the proposition that that is merely by way of example, and it does not affect the validity of the argument. The geometric form in which the argument is cast is also characteristic of ancient Greek mathematics; but that, too, is an inessential detail. More importantly, Euclid

© Springer International Publishing Switzerland 2015
J.W. Dawson, Jr., *Why Prove it Again?*, DOI 10.1007/978-3-319-17368-9_7

does not assume *anything* about the 'assigned' numbers except that they are primes. They need not be consecutive, nor even distinct, and there is certainly no assumption that they constitute *all* the primes. Thus, Euclid's proof is **not indirect**, as is often claimed. It is true that at one point Euclid assumes, by *reductio*, that G is equal to one of A, B, or C. But that *reductio* can easily be eliminated. Thus the argument may be paraphrased in modern terms as follows:

Let $p_1, p_2, \ldots, p_k$ be any prime numbers, and let L be their least common multiple. Then $N = L + 1$ is either a prime q itself or, by proposition VII,31, is divisible by some prime q. In the first case, $N = q$ is a prime distinct from $p_1, p_2, \ldots, p_k$, since it is larger than their least common multiple L (a justification that Euclid neglects to give). In the second case, q cannot be equal to any of $p_1, p_2, \ldots, p_k$, since N, when divided by any one of those numbers, leaves a remainder of 1.

The assumption that L is the *least* common multiple of $p_1, p_2, \ldots, p_k$ is unnecessary; any common multiple of $p_1, p_2, \ldots, p_k$ will do. That assumption does, however, reduce the size of the upper bound for q. If $p_1, p_2, \ldots, p_k$ are all distinct, then their least common multiple L is $p_1 p_2 \cdots p_k$, and if $p_1 < p_2 < \cdots < p_k$ are *all* the primes up to p_k, then $p_k < q$. If the sequence $p_1 < p_2 < \cdots < p_k$ does *not* include all the primes up to p_k, then in general one can only say that $q \leq p_1 p_2 \cdots p_k + 1$. For example, if $k = 2$, $p_1 = 3$ and $p_2 = 5$, then $N = 16$ and $q = 2$; and even if $p_1, p_2, \ldots, p_k$ are all the primes up to p_k, $N = L + 1$ need not be prime. The first counterexample occurs for $k = 6$, when $N = 2 \cdot 3 \cdot 5 \cdot 7 \cdot 11 \cdot 13 + 1 = 30,031 = 59 \cdot 509$.

A drawback to Euclid's construction is that the number $N = p_1 p_2 \cdots p_k + 1$ grows large rapidly, so that factoring it into primes to find a particular new prime q may be very time-consuming.[1] A generalization of Euclid's method that produces smaller values for q was given by Stieltjes in the first chapter of a projected text on the theory of numbers (Stieltjes 1890, p. 14).[2] His simple construction is based on proposition VII,28 of the *Elements*, which states that if A and B are relatively prime, then $A + B$ is relatively prime to both A and B, and conversely, if $A + B$ is relatively prime to one of A, B it is relatively prime to the other too.[3] Given that result of Euclid, Stieltjes considers rewriting the product $p_1 p_2 \cdots p_k$ of distinct prime numbers $p_1, p_2, \ldots, p_k$ as a product AB in any way whatever. By the Fundamental Theorem of Arithmetic (proven by Stieltjes on the previous page), A and B must be relatively prime, so no prime factor of $A + B$ can equal any of the p_i. Euclid's construction is the special case that results by taking either A or B equal to 1. The advantage of Stieltjes's method is readily apparent, even for $k = 3$. For then

---

[1] If $p_1, p_2, \ldots, p_k$ are the first k primes, then since $p_1 p_2 \ldots p_k + 1 < 2 \cdot p_1 p_2 \ldots p_k$, one can show by induction that $p_{k+1} \leq 2^{(2^k)}$.

[2] In 2008, essentially the same proof appeared in the *American Mathematical Monthly* (Cusumano et al. 2008), without reference to Stieltjes.

[3] Euclid's proof is by *reductio*, but the contrapositive, that if D divides any two of A, B, $A + B$ it must divide the third as well, is easily proven directly.

$AB = p_1 p_2 p_3 = 2 \cdot 3 \cdot 5 = 30$, so Euclid's construction yields $q = 31$, whereas if A is taken to be 2, 3 or 5, Stieltjes's construction yields that $A + B = 17, 13$ or 11, respectively.

The infinite sequence of numbers generated by Euclid's construction can be described recursively by the conditions $s(1) = 2$ and $s_{n+1} = s_1 s_2 \cdots s_n + 1$. The terms of that sequence are distinct integers greater than 1 that are pairwise relatively prime, and those conditions suffice to demonstrate the infinitude of the primes. (For if $a_1, a_2, \ldots$ is such a sequence, let p_i be any prime factor of a_i, for each integer $i \geq 1$.) Alternatively, one may consider the sequence defined for each positive integer n by $p_n =$ the least prime that divides $n! + 1$. Then clearly $p_n > n$ for every n. The sequence so defined is not one-to-one (for example, $p_5 = p_{10} = 11$), but as noted in Narkiewicz (2000), it does contain *all* primes; for by Wilson's Theorem, for every prime $q, (q-1)! \equiv -1 \pmod{q}$, so $p_{q-1} = q$.

The Fermat numbers $2^{2^n} + 1$ are an example of a sequence other than Euclid's that satisfies the aforementioned conditions. (A proof that they are pairwise relatively prime is given in Aigner and Ziegler 2000, pp. 3–4.) But use of the Fermat numbers to prove the infinitude of the primes seems *ad hoc* compared to Euclid's very natural construction, so that proof seems less desirable from a pedagogical standpoint. A different argument involving powers of 2 does, however, yield a simple alternative proof. For

Theorem: If p is a prime, then any prime factor q of $2^p - 1$ must be greater than p.

Proof: To say that q is a factor of $2^p - 1$ is equivalent to saying that $2^p \equiv 1 \pmod{q}$. In particular, $q \neq 2$, so if q is prime, $2^{q-1} \equiv 1 \pmod{q}$ also, by Fermat's Little Theorem. The division algorithm implies that if m is the least power of 2 for which $2^m \equiv 1 \pmod{q}$, then m must divide any n for which $2^n \equiv 1 \pmod{q}$, so since p is prime and $2^p \equiv 2^{q-1} \equiv 1 \pmod{q}$, p must divide $q - 1$. Therefore $q > p$.

All the proofs considered so far are constructive, in the sense that they provide an upper bound on the size of a prime distinct from those already known. A very strong result of that form is Bertrand's Postulate, which states that given any natural number $n \geq 1$, there is a prime q satisfying $n < q \leq 2n$. Bertrand's Postulate obviously implies the infinitude of the primes, but it is so much more difficult to prove that we do not consider it here as an alternative proof of the latter.[4]

Another approach to proving the infinitude of the primes is to use a counting argument to obtain a *lower* bound on the number of primes less than or equal to an integer N, and then show that that lower bound must approach infinity as N does.[5] A simple example of that approach is the following proof from Hardy and Wright

---

[4]But see Aigner and Ziegler (2000) for the ingenious, relatively short proof of Bertrand's Postulate given in 1932 by Paul Erdős (his first publication).

[5]As noted above in footnote 1, Euclid's proof can be analyzed to yield such a lower bound as well.

(1938), which can be understood without knowledge of calculus; it suffices merely to know that $\log_2 x$ denotes the exponent to which 2 must be raised to yield x, since it is clear that that exponent must grow arbitrarily large as x does.

Theorem: For $N \geq 1$, let k be the number of primes less than or equal to N. Then $k \geq \dfrac{1}{2} \log_2 N$.

Proof: The strategy is to estimate how many positive integers $n \leq N$ are divisible only by primes less than or equal to N. By the Fundamental Theorem of Arithmetic, the prime factorization of any such n is unique, say $n = p_1^{r_1} p_2^{r_2} \cdots p_k^{r_k}$. Let $s = p_{i_1} p_{i_2} \cdots p_{i_j}$, where $r_{i_1}, r_{i_2}, \ldots, r_{i_j}$ are all the *odd* exponents in that factorization of n, and let t be the product of all the remaining factors, so that $n = st$. The power of each prime factor in t must be *even*, so t can be rewritten as u^2. Therefore $n = su^2$, where s is square-free. Since, for each $1 \leq i \leq k$, p_i either does or does not occur in s, there are 2^k possible values for s; and since $n = su^2 \leq N, u^2 \leq N$, i.e., $u \leq \sqrt{N}$. There are thus at most $2^k \sqrt{N}$ integers less than or equal to N that are divisible by the k primes less than or equal to N. But the prime factors of *every* $n \leq N$ must themselves be less than or equal to N, so $2^k \sqrt{N} \geq N$. That is, $2^k \geq \sqrt{N}$, or $k \geq \dfrac{1}{2} \log_2 N$.

Another, more sophisticated way of establishing a lower bound for the number $\pi(x)$ of primes less than or equal to x may be traced back to Euler, who in 1737, in his *Introductio in analysin infinitorum*, claimed that

$$\sum_{i=1}^{\infty} \frac{1}{n} = \prod_{p \ prime} \frac{1}{1 - 1/p}.$$

Of course, either by invoking the integral test (Figure 7.1) or, more simply, by observing that $1 + 1/2 + 1/3 + \cdots + 1/2^n \geq 1 + n/2$, the harmonic series is seen to diverge.

Fig. 7.1 Divergence of the harmonic series

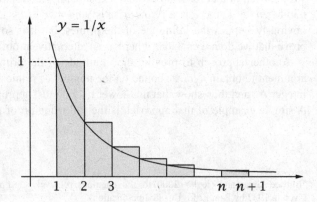

But how is an 'equation' involving a divergent series to be interpreted? That defect in the argument was repaired much later by Leopold Kronecker, who, in the first volume of his *Vorlesung über Zahlentheorie*, gave a correct proof that the Fundamental Theorem of Arithmetic can be expressed in terms of infinite sums and products.[6]

Proof (adapted from that by Euler): Let $\mathbb{P}$ denote the set of primes and N_x denote the set of natural numbers all of whose prime divisors are less than or equal to x. Then by the Fundamental Theorem of Arithmetic, every $m \in N_x$ has a unique representation as a product $\prod p_i^{r_i}$ of powers of primes $p_i \leq x$, so

$$\sum_{m \in N_x} \frac{1}{m} = \prod_{\substack{p \in \mathbb{P} \\ p \leq x}} \left(\sum_{k \geq 0} \frac{1}{p^k} \right) = \prod_{\substack{p \in \mathbb{P} \\ p \leq x}} \frac{1}{1 - 1/p} = \prod_{\substack{p \in \mathbb{P} \\ p \leq x}} \frac{p}{p-1} = \prod_{i=1}^{\pi(x)} \frac{p_i}{p_i - 1} .$$

If there were a largest prime, say $q = p_k$, then setting $x = q$ would give $N_x = \mathbb{N}$ and $\pi(x) = k$, so the last equation would reduce to

$$\sum_{m=1}^{\infty} \frac{1}{m} = \prod_{i=1}^{k} \frac{p_i}{p_i - 1} .$$

The right side would then be finite, contradicting the divergence of the harmonic series. That suffices to show the infinitude of the primes, but the proof can be extended to get a lower bound on $\pi(x)$ as follows:

If $x = 2^n$, then $\{1, 2, \ldots, 2^n\} \subseteq N_x$, so

$$1 + \frac{n}{2} \leq \sum_{i=1}^{2^n} \frac{1}{i} \leq \sum_{m \in N_x} \frac{1}{m} = \prod_{i=1}^{\pi(2^n)} \frac{p_i}{p_i - 1} ,$$

and since $p_i \geq i + 1$ for every i,

$$\frac{p_i}{p_i - 1} = 1 + \frac{1}{p_i - 1} \leq 1 + \frac{1}{i} = \frac{i+1}{i} .$$

Therefore

$$1 + \frac{n}{2} \leq \prod_{i=1}^{\pi(2^n)} \frac{i+1}{i} = \pi(2^n) + 1 .$$

----

[6]Kronecker replaced n and p by n^A and p^A, respectively, with $A > 1$, thereby obtaining a convergent series.

That is, for any positive integer n there are at least $n/2$ primes less than or equal to 2^n. (Note that this estimate agrees with that given in footnote 1 and with that given by the Hardy/Wright approach.)

The last proof to be considered here is an indirect one that uses topological concepts. It is due to Harry (= Hillel) Furstenberg, who published it in 1955 while an undergraduate student at Yeshiva University (Furstenberg 1955 reproduced in Aigner and Ziegler 2000, p. 5).

Proof (Furstenberg): Let $\mathbb{Z}$ denote the set of all integers and let $C(a, b)$ denote the congruence class $\{x \in \mathbb{Z} \,|\, x \equiv a \mod b\}$, whose members are the elements of the arithmetic progression $\{a + nb \,|\, n \in \mathbb{Z}\}$. The classes $C(a, b)$ form a neighborhood base for a topology on $\mathbb{Z}$. That is, a set O of integers is defined to be *open* if it is either empty or contains a neighborhood $C(a, b)$ of each point within it. Then unions of open sets are open, and if a is a point in $O_1 \cap O_2$ with $C(a, b_1)$ contained in O_1 and $C(a, b_2)$ contained in O_2, then $C(a, b_1 b_2)$ is contained in $O_1 \cap O_2$. Any nonempty open set must be infinite, and each $C(a, b)$ is closed as well as open, since it is the complement in $\mathbb{Z}$ of

$$\bigcup_{i=1}^{b-1} C(a + i, b).$$

Also, the complement in $\mathbb{Z}$ of $\{-1, 1\}$ is $\bigcup_{p \in \mathbb{P}} C(0, p)$, since very integer except -1 and 1 must be divisible by some prime.

So if $\mathbb{P}$ is assumed to be finite,

$$\{-1, 1\} = \bigcap_{p \in \mathbb{P}} C(0, p)$$

is a finite intersection of open sets. As such, $\{-1, 1\}$ must be a nonempty open set. But as noted earlier, all such sets are infinite. Hence by *reductio*, $\mathbb{P}$ must be infinite.

At first glance, Furstenberg's proof appears mysterious. Unlike the other proofs presented here, it seems to be entirely nonconstructive, yielding neither an upper nor lower bound on the size of $\pi(x)$; and the result seems to fall out unexpectedly. But (tellingly) no topological *theorems* are invoked in the proof, only topological *notions*; and when the topological terminology is unwound, the mystery is dispelled. Indeed, when examined more closely, the equation

$$\{-1, 1\} = \bigcap_{p \in \mathbb{P}} C(0, p)$$

says that 1 must be divisible by each of the (only) finitely many primes that were assumed to exist, when in fact it is divisible by none of them. The contradiction is then just that of Euclid's proof, when the latter is recast as a *reductio*: for since every integer is divisible by some prime, if there were only finitely many primes $p_1, p_2, \ldots, p_n$, then $p_1 p_2 \cdots p_n$ and $p_1 p_2 \cdots p_n + 1$ would necessarily share some common prime factor, which would then divide their difference, 1.

Furstenberg's proof is then simply a *reductio* version of Euclid's, cloaked in topological terminology.[7]

7.1 Summary

Despite universal agreement that Euclid's proof of the infinitude of the primes is an exemplar of simplicity and beauty, many alternative proofs have been given. Why?

The answer has primarily to do with *other applications* of the techniques used in the different proofs. Euclid's method, e.g., is easily modified to show that there are arbitrarily long gaps in the sequence of primes.[8] It is not suited, however, to proving Dirichlet's theorem on the infinitude of primes in arithmetic progressions, whereas Euler's approach is. Indeed, Euler's (attempted) proof of the infinitude of the primes was but one of many remarkable results that he obtained in Chapter XV of his *Introductio*, devoted to series that arise from products. Others include his proof that the series of reciprocals of the primes diverges, and his startling discovery that $\zeta(2) = \sum_{n=1}^{\infty} 1/n^2 = \pi^2/6$. Generalizations of the techniques he introduced there for manipulating series and products have become fundamental tools in analytic number theory.

Fermat's Little Theorem is another basic tool in number theory, applicable in many contexts, including the theory underlying public-key cryptography, so vital to today's computerized communications and banking.

Likewise, the key idea in the Hardy/Wright approach, that of representing any integer $n \geq 2$ in a unique way as a product st, where s is square-free, also has wider applications. In particular, it too can be used as a tool in proving the divergence of the series of prime reciprocals. (See Erdős 1938 and Niven 1971.)

References

Aigner, M., Ziegler, G.M.: Proofs from the Book, 2nd ed. Springer, Berlin (2000)

Cusumano, A., Dudley, U., Dodge, C: A variation on Euclid's proof of the infinitude of the primes. Amer. Math. Monthly **115**(7), 663 (2008)

Erdős, P.: Über die Reihe $\Sigma 1/p$. Math. Zutphen B **7**, 1–2 (1938)

Furstenberg, H.: On the infinitude of primes. Amer. Math. Monthly **62**(5), 353 (1955)

Hardy, G.H., Wright, G.M.: Introduction to the Theory of Numbers. Clarendon, Oxford (1938)

Hardy, M., Woodgold, C.: Prime simplicity. Math. Intelligencer **31**(4), 44–52 (2009)

Heath, T.: The Thirteen Books of Euclid's Elements (3 vols.). Dover, New York (1956)

Narkiewicz, W.: The Development of Prime Number Theory. Springer, Berlin (2000)

Niven, I.: A proof of the divergence of $\Sigma 1/p$. Amer. Math. Monthly **78**(3), 272–273 (1971)

Stieltjes, T.J.: Sur la théorie des nombres. Ann. fac. sciences Toulouse **IV**(1)(3), 1–103 (1890)

---

[7]Of course, in Euclid's proof one could just as well let $N = L - 1$ instead of $L + 1$.

[8]E.g., if $n \geq 2$, then none of the $n - 1$ numbers $n! + 2, n! + 3, \ldots, n! + n$ can be prime.

Chapter 8
The Fundamental Theorem of Algebra

The Fundamental Theorem of Algebra (stated below) provides an ideal case study for illustrating the roles of alternative proofs in mathematical practice. Like the Pythagorean Theorem, the Fundamental Theorem of Algebra has been proved in many different ways since its enunciation by Euler in 1739. Unlike the Pythagorean Theorem, however, early attempts to prove the Fundamental Theorem of Algebra are not shrouded in the mists of antiquity, so we know how the adequacy of those attempts was evaluated by mathematicians of the time. We can see how criticisms of earlier efforts to prove the theorem led to alternative proof strategies, and we can analyze why the proof given by Gauss in his 1799 inaugural dissertation was the first to be accorded general acceptance, though it too would later be deemed not fully rigorous.

As with the theorems considered in earlier chapters, besides questions of rigor there have been other impetuses for devising alternative proofs of the Fundamental Theorem of Algebra: issues of perspicuity, simplicity, generality, purity of method and constructivity have also been matters of concern; and in a pedagogical context, different proofs of the Fundamental Theorem have been employed as a vehicle for introducing a variety of topics in higher-level mathematics (complex line integrals, field extensions, Galois theory, and notions from algebraic topology) in a text designed for a capstone course for senior mathematics majors (Fine and Rosenberger 1997).

8.1 Alternative formulations of the theorem

In its earliest and simplest form, the Fundamental Theorem of Algebra was the conjecture that every polynomial with real coefficients can be expressed as a product of linear and quadratic polynomials with real coefficients. The question whether that is so arose in connection with Leibniz's attempts to integrate functions by the method of partial fractions, and Leibniz himself believed the conjecture to be false.

© Springer International Publishing Switzerland 2015
J.W. Dawson, Jr., *Why Prove it Again?*, DOI 10.1007/978-3-319-17368-9_8

Euler, however, showed that putative counterexamples put forward by Leibniz and by Nikolaus Bernoulli did, in fact, possess factorizations of the stated form, and it was he, in a letter to Bernoulli of 1 October 1742, who first asserted the truth of the statement. Two months later, however, in a letter of 15 December to his friend Goldbach, he confessed that he was unable to produce a fully satisfactory proof of the theorem.[1]

Today the Fundamental Theorem of Algebra is more often stated in the form "Every polynomial $p(X)$ of degree n with complex coefficients possesses exactly n complex roots, counting multiplicities." The equivalence of that statement with the one given above rests not only upon recognizing complex numbers as meaningful entities, but upon the quadratic formula (which shows how to express any quadratic polynomial as a product of two complex linear factors), upon the factor theorem of Descartes (that a is a root of a polynomial $p(X)$ if and only if $X - a$ is a factor of $p(X)$), and upon the observation (made by Bombelli around 1560, and again by Euler in his 1742 letter to Goldbach) that the complex roots of any polynomial with real coefficients always occur in conjugate pairs, so that the product of the corresponding linear factors guaranteed by Descartes's theorem is a *real* quadratic polynomial.

Consideration of the properties of complex conjugates shows that if $p(X)$ is a polynomial with complex coefficients and $\overline{p}(X)$ is the polynomial whose coefficients are the conjugates of those of $p(X)$, then the product $p(X)\overline{p}(X)$ is a polynomial with *real* coefficients. If z_0 is a complex root of $p(X)\overline{p}(X)$, then it must either be a root of $p(X)$ or of $\overline{p}(X)$; and in the latter case, again using the overbar to denote complex conjugation, $\overline{\overline{p}(z_0)} = \overline{\overline{p}}(\overline{z_0}) = p(\overline{z_0}) = 0$, so $\overline{z_0}$ is a root of $p(X)$. To prove the Fundamental Theorem it therefore suffices to establish it for polynomials $p(X)$ whose coefficients are real numbers.

8.2 Early attempts to prove the theorem

The task of showing that every polynomial with real coefficients possesses at least one complex root involves two separate aspects: showing (1) that a root *of some definite sort* exists, and (2) that any such root must in fact be of the form $a + bi$ (in modern terms, proving the existence of a splitting field K over $\mathbb{R}$ for $p(X)$, and then showing that K must be isomorphic to $\mathbb{C}$). Before Gauss, however, those who endeavored to prove the existence of complex roots either explicitly *assumed* (1) to be true (in part, perhaps, because it was believed that formulas for the roots of polynomials of degree ≥ 5 similar to those obtained by Ferro, Tartaglia, Ferrari, and

---

[1]The dates given here for Euler's letters are based on the account in Kline (1972), pp. 597–598. They disagree with those given in Remmert (1990), which are inconsistent with one another.

Bombelli for cubic and quartic polynomials would eventually be found[2]), or else unwittingly employed arguments that smuggled in that assumption. Their efforts focused instead on establishing (2). But, as Gauss trenchantly observed in the critique of prior proof attempts that he gave in his dissertation, there was need not only to *justify* the existence of roots, but, if algebraic operations were to be applied to them, to characterize their structure; for it made no sense to attempt to manipulate hypothetical quantities that were mere "shadows of shadows."

The remainder of this section is devoted to outlining the strategies employed by D'Alembert, Euler, Lagrange, and Laplace in their attempted proofs of the Fundamental Theorem (published in 1746, 1749, 1772 and 1795, respectively) and to analyzing the deficiencies in their arguments.

d'Alembert's 'proof': In his memoir 'Recherches sur le calcul intégral' (d'Alembert 1746), Jean Le Rond d'Alembert is generally credited with having made the first serious attempt to prove the Fundamental Theorem of Algebra. The memoir was apparently hastily written, however, and is not notable for its clarity. Indeed, there is marked disparity among the descriptions given by modern commentators both of the mechanics of d'Alembert's argument and of the extent of its deficiencies. (My own reading of d'Alembert's text is in accord with the descriptions of it in Gilain (1991) and Baltus (2004), but at variance with that in Remmert (1990).) d'Alembert began by noting that if $p(X) = X^m + c_{m-1}X^{m-1} + \cdots + c_1 X + c_0$ is a monic polynomial with real coefficients of degree $m \geq 1$, then $p(0) = 0$ if $c_0 = 0$. He then replaced the constant term c_0 by the *parameter* z and set the resulting function $F(X, z)$ equal to 0, so that the Fundamental Theorem became the statement that for any real value of z, there is a (possibly complex) value x for which $F(x, z) = 0$. To establish that, d'Alembert first claimed that for any real number z_0, if (x_0, z_0) is a point for which $F(x_0, z_0) = 0$ — in particular, if $x_0 = z_0 = 0$ — then for all real z sufficiently close to z_0, there is a complex value x for which $F(x, z) = 0$. He then went on to claim that for any real value z^* of z, an overlapping chain of discs can be found, starting at $(0, 0)$, yielding a sequence of points (x_n, z_n) such that $F(x_n, z_n) = 0$ for each n, the values x_n converge to a complex number x^* and the (real) values z_n to z^*, with $F(x^*, z^*) = 0$.

To establish the first claim d'Alembert alleged, without proof, that if

$$F(x_0, z_0) = 0,$$

then for all z sufficiently close to z_0, there is a natural number q and a convergent series of fractional powers of $z - z_0$ such that

---

[2]The impossibility of expressing the roots of arbitrary polynomials of degree ≥ 5 in terms of radicals was finally established by Abel in 1826.

$$x = x_0 + \sum_{i=1}^{\infty} c_i (z - z_0)^{i/q}$$

satisfies $F(x, z_0) = 0$. More than a century later that fact was finally proved by Victor Puiseaux, as a *consequence* of the Fundamental Theorem (which by then had been rigorously proved by other means); so d'Alembert's argument was circular. There were difficulties, too, with his second claim: What ensures that the radii of the discs are such that the z_n converge to z^* ? And even if they do, why does the fact that $F(x_n, z_n) = 0$ for each n entail that $F(x^*, z^*) = 0$? Those and other criticisms were lodged against d'Alembert's argument by Gauss, who nevertheless thought that it might be possible to repair its defects. But he and others chose instead to seek different ways to establish the Fundamental Theorem.[3]

Euler's attack on the theorem: Three years after the appearance of d'Alembert's memoir, Euler attempted to prove the Fundamental Theorem in its original formulation. By invoking the fact that a real polynomial of odd degree must have a real root (a consequence of the intermediate-value theorem, a principle generally accepted at the time, but first rigorously proved by Bolzano around 1816), he argued that a real quintic polynomial must have at least one real linear factor, and then went on to show how any real quartic polynomial could be expressed as a product of two real quadratic factors. Having thus established the truth of the Fundamental Theorem for polynomials of degree ≤ 5, he attempted to extend the proof to polynomials of higher degree, but was unable to do so.

At first glance it might appear that Euler had made but a minor advance beyond the work of Ferrari and Bombelli two centuries earlier, since their explicit formulas for the roots of real quartic polynomials, in which the complex roots occur in conjugate pairs, immediately entail that all such quartics can be factored into a product of real polynomials of degree at most 2. But in the formulas obtained by the Italians for the roots of cubic and quartic polynomials, complex numbers play an essential role. In its original form, however, the Fundamental Theorem makes no reference to complex numbers, so, as noted in Remmert (1990), p. 117, their employment in proofs thereof appears to invoke a *deus ex machina*. Euler's method of factoring quartics, however, made no use of complex numbers, so from the standpoint of purity of method it was superior.

A detailed and very readable discussion of what Euler did in his paper Euler (1749) is given in Dunham (1991). Following the lead of the Italian school, Euler noted that any monic polynomial of degree 4 in the variable x with real coefficients can be converted into an equivalent quartic in the variable y that lacks a cubic term (via the substitution $x = y - c_3/4$, where c_3 is the coefficient of x^3 in the original polynomial). The factorization of the resulting quartic $y^4 + By^2 + Cy + D$ then

---

[3] An attempt to repair d'Alembert's proof is given in Baltus (2004), but that effort, too, appears to be flawed.

depends on the values of the coefficients B, C and D. If $C = 0$, the quartic is a quadratic in y^2, which, if $B^2 - 4D \geq 0$, factors as

$$\left[y^2 + \frac{B + \sqrt{B^2 - 4D}}{2} \right] \left[y^2 + \frac{B - \sqrt{B^2 - 4D}}{2} \right].$$

On the other hand, if $B^2 - 4D < 0$, then both $\sqrt{D} > 0$ and $\sqrt{2\sqrt{D} - B} > 0$, so

$$y^4 + By^2 + D = \left[y^2 + \sqrt{D} \right]^2 - \left[y\sqrt{2\sqrt{D} - B} \right]^2,$$

a difference of two squares that factors once again into the product of two quadratics. If $C \neq 0$, the absence of the y^3 term in $y^4 + By^2 + Cy + D$ implies that any factorization of that quartic into quadratic factors must be of the form $(y^2 + uy + \alpha)(y^2 - uy + \beta)$ for some constants u, α and β. Expanding that product, setting it equal to $y^4 + By^2 + Cy + D$ and equating coefficients of like powers of y, Euler obtained three equations in the unknowns u, α and β, from which after further algebra he deduced the equation $u^6 + 2Bu^4 + (B^2 - 4D)u^2 - C^2 = 0$, in which all the powers of u are even. The graph of $Y = u^6 + 2Bu^4 + (B^2 - 4D)u^2 - C^2$ is therefore symmetric about the Y-axis, Y approaches $+\infty$ as u approaches $\pm\infty$, and $Y(0) < 0$ (see Figure 8.1), so by the intermediate-value principle, $Y = u^6 + 2Bu^4 + (B^2 - 4D)u^2 - C^2$ must have real roots $\pm u_0$, either of which can be substituted back into the earlier equations to find real values for α and β.

To extend to polynomials of arbitrary degree, Euler noted that by multiplying, if necessary, by some positive integral power of X, any polynomial $p(X)$ of degree

Fig. 8.1 Graph of a sixth-degree polynomial with no odd powers

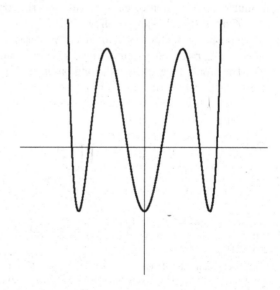

d could be converted into a polynomial $q(X)$ of degree 2^m, for some $m > 0$. He then attempted to mimic the procedure he had employed for factoring quartics. A direct approach led to systems of equations too complex to allow derivation of an equation for u, but he showed that an alternative approach, avoiding the need to find an explicit equation for u_0, was also possible in the quartic case. However, to extend that approach to polynomials of degree 2^m for $m > 2$ it was necessary to *assume* that $2m$ roots of some sort existed; and without specifying the nature of those roots, Euler's attempt to show that algebraic combinations of them would yield *real* coefficients for the putative factors of $q(X)$ was doomed to failure (as Gauss was to point out).

Lagrange's improvement of Euler's argument: In a long and important paper that appeared in 1770/1771,[4] Joseph Louis Lagrange investigated the properties of symmetric polynomials and established the result now known as the fundamental theorem about them: that any polynomial $S(X_1, \ldots, X_n)$ symmetric in $X_1, \ldots, X_n$ has a unique representation as a polynomial $P(s_1, \ldots, s_n)$, where $s_1, \ldots, s_n$ are the elementary symmetric polynomials in $X_1, \ldots, X_n$. Using that result, and assuming that a real polynomial $p(X)$ of degree 2^m had 2^m roots that could be manipulated like ordinary real numbers (in modern terms, that $p(X)$ had roots in some field extending $\mathbb{R}$), Lagrange was able to establish (even to Gauss's satisfaction) that the factors Euler had sought for $p(X)$ would indeed have real coefficients. Only the justification for the existence of such roots remained to be proven.

Laplace's proof: Under the same basic assumptions that Euler and Lagrange had made (the existence of a splitting field and the intermediate-value principle), together with DeMoivre's theorem on roots of complex numbers, proved earlier in the eighteenth century, Pierre Simon de Laplace employed Lagrange's theorem on symmetric polynomials to prove the Fundamental Theorem of Algebra in its second formulation. A version of his proof in modern terminology, as given in Remmert (1990), pp. 120–122, goes as follows.[5]

Suppose a monic polynomial $p(X)$ of degree $n \geq 1$ with real coefficients has roots $r_1, \ldots, r_n$ in some splitting field F over $\mathbb{R}$, and rewrite n as $2^m q$, where q is odd. The proof proceeds by induction on m. If $m = 0$, $p(X)$ has a real root by the intermediate-value principle. For $m \geq 1$, suppose that every polynomial of degree $n = 2^k q$ with $k < m$ has a complex root. Laplace then considered the symmetric polynomials over F given by

$$L_t(X) = \prod_{1 \leq i < j \leq n} (X - r_i - r_j - t r_i r_j),$$

---

[4]"Réflexions sur la résolution algébrique des équations," reprinted in *Oeuvres de Lagrange* III, 205–421)

[5]Full background details can be found in Chapter 6 of Fine and Rosenberger (1997), where, however, the strategy underlying the proof is not credited to Laplace.

for each positive integer t. By the Fundamental Theorem on Symmetric Polynomials, each L_t, when written as a polynomial in powers of X, has coefficients that are elementary symmetric polynomials in the roots of the *real* polynomial $p(X)$. But the coefficient of each power X^j in $p(X)$ *is* just $(-1)^j s_j(r_1, \ldots, r_n)$, where $s_j(r_1, \ldots, r_n)$ is the jth elementary symmetric polynomial in $r_1, \ldots, r_n$; so each coefficient of $L_t(X)$ is a real number. Moreover, each $L_t(X)$ has degree

$$\binom{n}{2} = \binom{2^m q}{2} = 2^{m-1} q [2^m q - 1],$$

where $q[2^m q - 1]$ is odd. By the induction hypothesis, each $L_t(X)$ thus has a complex root c_t, so for some pair (r_i, r_j) with $1 \le i < j \le n, c_t = r_i + r_j + t r_i r_j$; and since there are infinitely many integers t but only finitely many pairs (r_i, r_j) with $1 \le i < j \le n$, there must be *distinct* integers t_1 and t_2 such that for the *same i* and j, $c_{t_1} = r_i + r_j + t_1 r_i r_j$ and $c_{t_2} = r_i + r_j + t_2 r_i r_j$ are both complex numbers. The difference $c_{t_1} - c_{t_2} = (t_1 - t_2) r_i r_j$ is then also a complex number, whence so is $r_i r_j$. Therefore $c_{t_1} - t_1 r_i r_j = r_i + r_j$ is a complex number as well. So by DeMoivre's theorem and the quadratic formula, the roots of the polynomial $X^2 - (r_i + r_j)X + r_i r_j = (X - r_i)(X - r_j)$, that is, the roots r_i and r_j of the original polynomial $p(X)$, must be complex numbers.   q.e.d.

8.3 Gauss's first proof

Gauss's doctoral dissertation, submitted to the University of Helmstedt in 1799 and written in Latin, was entitled *Demonstratio nova theorematis omnem functionem algebraicam rationalem integram unius variabilis in factores reales primi vel secundi gradus resolvi posse*[6] — that is, "New proof of the theorem that every rational integral algebraic function [i.e, polynomial] of one variable can be resolved into real factors of first or second degree." Gauss thus stated the Fundamental Theorem in its original formulation, and declared his aim to be that of giving "a new and stronger proof" of that result. On the second page of the dissertation he noted the equivalent formulation in terms of complex roots, but stated that he would eschew the use of complex numbers in his demonstration. He went on to criticize earlier proofs, all of which he faulted for presuming without justification that a polynomial of degree m must possess m roots of some (unspecified) sort. To avoid that presumption, he gave a geometric argument to establish the desired factorization.

---

[6]Reprinted in Gauss's *Werke* III, 1-30. The discussion here is based on the German translation by E. Netto in Gauss (1890).

Outline of proof: Gauss begins by proving two lemmas.

Lemma 1: If m is any positive integer, then $x^2 - 2r\cos\phi x + r^2$ is a factor of $\sin\phi x^m - r^{m-1}\sin(m\phi)x + r^m\sin(m-1)\phi$.

(The latter expression is 0 if $m = 1$. If $m = 2$, the other factor is $\sin\phi$, and if $m > 2$, $\sum_{i=1}^{m-1}\sin(i\phi)r^{i-1}x^{m-i-1}$ is the other factor.)

Lemma 2: If r and ϕ satisfy the equations

$$(8) \quad r^m\cos(m\phi) + Ar^{m-1}\cos(m-1)\phi + Br^{m-2}\cos(m-2)\phi + \ldots$$
$$+ Kr^2\cos(2\phi) + Lr\cos\phi + M = 0$$

and

$$(9) \quad r^m\sin(m\phi) + Ar^{m-1}\sin(m-1)\phi + Br^{m-2}\sin(m-2)\phi + \ldots$$
$$+ Kr^2\sin(2\phi) + Lr\sin\phi = 0,$$

then the expression $x^m + Ax^{m-1} + Bx^{m-2} + \cdots + Kx^2 + Lx + M$ has the factor $x - r\cos\phi$ if $r\sin\phi = 0$ and the factor $x^2 - (2r\cos\phi)x + r^2$ if $r\sin\phi \neq 0$.
(Gauss notes that complex numbers are usually invoked to prove Lemma 2, but he gives an alternative proof that avoids them, based on Lemma 1.)

To prove the Fundamental Theorem it therefore suffices to show that r and ϕ can be found that satisfy the two equations of Lemma 2.[7]
     Toward that end, Gauss considers the surfaces generated by the functions

$$T = r^m\sin(m\phi) + Ar^{m-1}\sin(m-1)\phi + Br^{m-2}\sin(m-2)\phi + \ldots$$
$$+ Kr^2\sin(2\phi) + Lr\sin\phi$$

and

$$U = r^m\cos(m\phi) + Ar^{m-1}\cos(m-1)\phi + Br^{m-2}\cos(m-2)\phi + \ldots$$
$$+ Kr^2\cos(2\phi) + Lr\cos\phi + M$$

---

[7]The connection with the complex formulation of the theorem is readily seen, since by DeMoivre's Theorem, if the variable X is written in polar form as $X = r(\cos\phi + i\sin\phi)$, the left members of those equations are just the real and imaginary parts of the expression $X^m + Ax^{m-1} + Bx^{m-2} + \cdots + Kx^2 + Lx + M$.

Fig. 8.2 Alternating T- and U-arcs

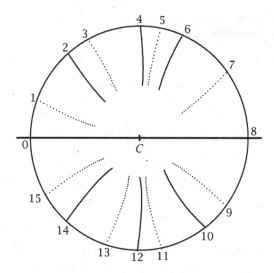

above and below the (r, ϕ)-plane and the traces in that plane where those surfaces intersect it.[8] The problem then becomes that of showing that there is at least one point in the (r, ϕ)-plane where the T-trace and the U-trace themselves intersect.

Further analysis shows that the T- and U-traces each contain $2m$ arcs that extend to infinity. Two arcs of the T-trace join to form the horizontal axis. The other arcs are each asymptotic to lines where $\sin(m\phi) = 0$, that is, to lines through the origin that are inclined to the axis at one of the angles $k\pi/m$, for $0 < k < m$. The arcs of the U-trace are likewise asymptotic to lines where $\cos(m\phi) = 0$, that is, to lines through the origin that are inclined to the axis at one of the angles $(2k - 1)\pi/2m$, for $0 < k \leq m$. Accordingly, those arcs will intersect a circle of sufficiently large radius at $2m$ points, which divide its circumference into $2m$ intervals in which T is alternately positive and negative. Moreover, those points are alternately one where a T-arc intersects the circle and one where a U-arc does. (See Figure 8.2, based on Gauss's own illustration for the quartic polynomial $X^4 - 2X^2 + 3X + 10$. The solid curves there represent the T-arcs and the dotted ones the U-arcs.)

Assuming that for at least one k the arc of the T-trace that intersects the circle at point k and the arc of the T-trace that intersects the circle at point $k + 2$ are both part of the *same* continuous T-branch, and likewise that the arc of the U-trace that intersects the circle at point $k + 1$ and the arc of the U-trace that intersects the circle at point $k + 3$ are both part of the same continuous U-branch, then that U-branch,

---

[8] Since the leading terms of T and U dominate the others and can be made positive or negative by appropriate choice of the angle ϕ, it is clear by continuity that both surfaces do intersect the (r, ϕ)-plane.

Fig. 8.3 Intersecting T- and
U-branches

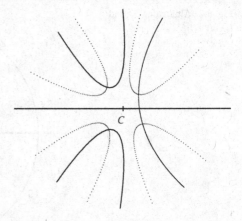

in passing from one intersection point where T is positive to another where it is
negative, must at some point within the circle cross that T-branch. (See Figure 8.3,
also taken from Gauss's original text.) The theorem would thereby be proved.

Gauss declared that the assumption involved could be justified in "many different
ways," one of which he endeavored to outline. But ultimately he had to rely on
his geometric intuition that "an algebraic curve can neither suddenly end abruptly
... nor lose itself, so to speak, ... after an infinity of circuits (as in the case of
a logarithmic spiral)" — a fact that he believed could "be taken as [having been]
sufficiently securely established" (Gauss 1890, footnote to p. 33).

Gauss's contemporaries evidently agreed, for they found no fault with his proof.
Only much later — long after Bolzano's proof of the intermediate-value theorem,
and after Kronecker, Dedekind and others, by relatively straightforward means, had
shown how to construct splitting fields and thereby justified the earlier proofs of
the Fundamental Theorem that had relied on those facts — did mathematicians
come to regard the principle that Gauss had relied on (that a non-compact branch
of an algebraic curve that enters a bounded space must eventually emerge from
it) as a statement (like the Jordan Curve Theorem) that required more rigorous
demonstration. The principle was finally proved rigorously by Alexander Ostrowski
in 1920, using sophisticated topological notions. (See Ostrowski 1983.)

Fifty years after his receipt of the doctorate, Gauss gave another proof of the
Fundamental Theorem (his fourth) that was a minor variant of the one given in
his dissertation.[9] Since (as he remarked) complex numbers had by then come to
be generally accepted by the mathematical community (in large part due to the
Fundamental Theorem itself), he felt free to employ them in the revised version of
his proof; and there he also allowed the polynomials to have complex coefficients.

---

[9]A detailed exposition of a modernized version of Gauss's fourth proof is given in Fine and
Rosenberger (1997), pp. 182–186.

8.4 Argand's proof

The assertion that every nonconstant polynomial with complex coefficients must have a complex root was first made not by Gauss, but by Jean Robert Argand, a Paris bookkeeper who introduced the planar representation of complex numbers now named after him;[10] and in 1814 Argand gave his own, very different proof of that result (Argand 1814), for whose understanding nothing beyond some basic knowledge of advanced calculus is required.

Specifically, Argand's proof depends on knowing (1) that polynomials are continuous functions; (2) that every continuous function defined on a closed disc $|z| \leq R$ assumes a minimum in that disc; and (3) that every complex number has a kth root for each integer $k > 1$ (an immediate consequence of DeMoivre's Theorem).[11] The proof then proceeds as follows:

Given a polynomial $p(z) = a_n z^n + a_{n-1} z^{n-1} + \cdots + a_1 z + a_0$ with $n \geq 1$, $|p(z)|$ approaches ∞ as $|z|$ does, so for any positive constant C, there is an $R > 0$ such that $|p(z)| > C$ for $|z| > R$. Taking $C = \inf_{z \in \mathbb{C}} |p(z)|$, it follows that $\inf_{z \in \mathbb{C}} |p(z)| = \inf_{z \leq R} |p(z)|$, so by (2), $|p((z)|$ assumes a minimum for some z_0 with $|z_0| \leq R$. That $p(z_0) = 0$ then follows directly from

Argand's inequality: For any polynomial $p(z)$ of degree $n \geq 1$, if $p(z_0) \neq 0$ then there is a $z_1 \in \mathbb{C}$ for which $|p(z_1)| < |p(z_0)|$.

Sketch of proof: Since $p(z_0) \neq 0$, we can divide $p(z)$ by $|p(z_0)|$ to obtain a polynomial $q(z)$ of the same degree for which $q(z_0) = 1$; and if we define $h(z) = q(z + z_0)$, then $h(0) = 1$, so $h(z)$ may be written as

$$1 + b_1 z + b_2 z^2 + \cdots + b_n z^n = 1 + b_k z^k + b_{k+1} z^{k+1} + \cdots + b_n z^n$$
$$= 1 + b_k z^k + z^k [b_{k+1} z + \cdots + b_n z^{n-k}],$$

where k is the least index i for which $b_i \neq 0$. The expression in brackets is a polynomial that has $z = 0$ as a root, so it is continuous at $z = 0$ and therefore can be made arbitrarily small for z sufficiently close to 0. If r is any kth root of $-1/b_k$, the triangle inequality then shows that $|h(rt)| < 1$ for sufficiently small positive real numbers t. For any such t, setting $z_1 = z_0 + rt$ yields $|p(z_1)| < |p(z_0)|$.

---

[10]See the entry on Argand by Phillip Jones in the *Dictionary of Scientific Biography*, vol. 1, pp. 237–240.

[11]Argand assumed fact (2), which was not proved rigorously until later.

8.5   Gauss's second proof

Despite the acceptance of his 1799 proof by other mathematicians of his time, Gauss published a second proof in 1815[12] that was based on algebraic rather than geometric principles. In his opening remarks, he maintained that his first proof "probably [sic] leaves nothing more to be desired with respect to rigor or simplicity".[13] He deemed his new proof to be "no less rigorous" than the first, but he did not claim it was simpler (as it certainly was not). Why then did he offer it?

Perhaps, as the qualifying *wohl* might be taken to suggest, he did after all harbor some doubts about whether the geometric principles he had invoked in his first proof had been rigorously established; but at least ostensibly, he was concerned about purity of method.

Like the proofs of Laplace and Lagrange, Gauss's 1815 proof was by induction on the highest power of 2 dividing the degree of the polynomial $Y(x)$. But Gauss avoided having to assume that $Y(x)$ had any roots by working in terms of indeterminates $a_1, a_2, \ldots, a_n$. He defined various auxiliary polynomials symmetric in those indeterminates, and applied the Fundamental Theorem on Symmetric Polynomials to each.[14]

The very long proof is divided into twenty numbered sections. To make the argument self-contained, Gauss first established a number of preliminary results: Given two polynomials $Y_1(x)$ and $Y_2(x)$, whose coefficients might include other indeterminates in addition to x, he defined their greatest common divisor and used the Euclidean algorithm to show that the g.c.d. must be a linear combination of $Y_1(x)$ and $Y_2(x)$, so that, in particular, if $Y_1(x)$ and $Y_2(x)$ have no common divisor of positive degree, there must be polynomials $Z_1(x)$ and $Z_2(x)$ such that $Z_1(x)Y_1(x) + Z_2(x)Y_2(x) = 1$, and conversely. In section 3, for any positive integer m he defined the elementary symmetric polynomials in the indeterminates $a_1, a_2, \ldots, a_m$ and noted that any polynomial function of *them* (again, possibly containing other indeterminates as well) must also be symmetric in $a_1, a_2, \ldots, a_m$. Section 4 was devoted to proving the converse (the fundamental theorem), and section 5 to establishing the uniqueness of that representation.

Then, for any integer $m \geq 1$, he defined π_m to be the symmetric polynomial in the indeterminates $a_1, a_2, \ldots, a_m$ given by

$$(10) \qquad \pi_m = \prod_{\substack{1 \leq i, j \leq m \\ i \neq j}} (a_i - a_j) = (-1)^{\binom{m}{2}} \prod_{\substack{1 \leq i, j \leq m \\ i < j}} (a_i - a_j)^2.$$

---

[12]Translated into German on pp. 37–60 of Gauss (1890).

[13]"*in Anbetracht der Strenge wie der Einfachheit wohl nichts zu wünschen übrig lässt*"

[14]In the description of Gauss's proof given below we depart from his notation in some respects. In particular, we use subscripts rather than primes or distinct letters to distinguish among different objects of the same type. For an overview in English that retains Gauss's notation, see Baltus (2000).

By the fundamental theorem, π_m can be represented as a polynomial in the elementary symmetric polynomials $s_1, s_2, \ldots s_m$ of the indeterminates $a_1, a_2, \ldots, a_m$. Let p_m be the polynomial in the indeterminates $i_1, i_2, \ldots, i_m$ obtained from the latter polynomial by replacing each occurrence of s_j therein by i_j, for $1 \leq j \leq m$.[15] Then, given a monic polynomial $Y(x) = x^m - C_1 x^{m-1} + C_2 x^{m-2} - \cdots \pm C_m$ of degree m with real coefficients $C_1, -C_2, \ldots, \pm C_m$, let P_Y be the real number obtained by substituting for each i_j in p_m the value C_j from Y (that is, P_Y is the value of $p_m(i_1, i_2, \ldots, i_m)$ at $(C_1, C_2, \ldots, C_m)$).

Sections 6–9 of Gauss's paper are devoted to proving

Lemma 1: If $Y'(x)$ is the (formal) derivative of $Y(x)$, then $Y(x)$ And $Y'(x)$ have a nonconstant common factor if and only if $P_Y = 0$.

Gauss began by noting that *if* $Y(x)$ could be decomposed into (not necessarily distinct) linear factors, say

$$Y(x) = (x - A_1)(x - A_2) \cdots (x - A_m),$$

then

$$Y(x) = x^m - \left(\sum_{1 \leq i \leq m} A_i \right) x^{m-1} + \left(\sum_{\substack{1 \leq i, j \leq m \\ i \neq j}} A_i A_j \right) x^{m-2} - \cdots \pm \prod_{1 \leq i \leq m} A_i \ ;$$

that is, for each $1 \leq i \leq m$, $C_i = s_i(A_1, A_2, \ldots, A_m)$. Hence

$$P_Y = p_m\left(s_1(A_1, A_2, \ldots, A_m), s_2(A_1, A_2, \ldots, A_m), \ldots, s_m(A_1, A_2, \ldots, A_m)\right)$$

$$= \prod_{\substack{1 \leq i, j \leq m \\ i \neq j}} (A_i - A_j),$$

so $P_Y = 0$ if and only if for some distinct i, j, $A_i = A_j$. It then follows directly from the product rule, applied to the factored form of $Y(x)$, that the latter condition holds if and only if $Y(x)$ and $Y'(x)$ have some factor $(x - A_i)$ in common.

To avoid circularity in proving the Fundamental Theorem, however, it was necessary to prove the Lemma without presuming that $Y(x)$ could be decomposed into linear factors. To do so, Gauss noted that any monic polynomial

$$Y(x) = x^m - C_1 x^{m-1} + C_2 x^{m-2} - \cdots \pm C_m$$

---

[15] Gauss called p_m the *determinant* of the polynomial $y(x) = x^m - i_1 x^{m-1} + i_2 x^{m-2} - \cdots \pm i_m$; today, it is called the *discriminant*. He described it as "that function of the indeterminates $i_1, \ldots, i_m$ that is transformed into the product of every pair of differences of distinct indeterminates $a_1, \ldots, a_m$ when each i_k is replaced by s_k." Note that p_m depends only on m.

in the indeterminate x could be regarded as a substitution instance of the polynomial

$$(11) \qquad y(x, i_1, \ldots, i_m) = x^m - i_1 x^{m-1} + i_2 x^{m-2} - \cdots \pm i_m,$$

in which the indeterminates $i_1, \ldots, i_m$ had been replaced by the real numbers $C_1, \ldots, C_m$. If, on the other hand, those indeterminates were replaced by the elementary functions $s_1, \ldots, s_m$ of the indeterminates $a_1, \ldots, a_m$, the resulting polynomial v could be written as

$$(12) \qquad v = \prod_{i=1}^{m} (x - a_i).$$

As a tool for proving the 'only if' direction of the Lemma, Gauss considered the polynomial in the indeterminates $x, a_1, a_2, \ldots, a_m$, symmetric in $a_1, a_2, \ldots, a_m$, given by

$$\rho = \pi_m \sum_{i=1}^{m} \prod_{\substack{1 \leq j \leq m \\ j \neq i}} \frac{(x - a_j)}{(a_i - a_j)^2}$$

(That ρ *is* a polynomial in $x, a_1, a_2, \ldots, a_m$ follows from the rightmost member of equation (10), which shows that the denominator in each summand of ρ divides π_m.) If $1 \leq k \leq m$ and $x = a_k$, then only one summand of ρ and one of the derivatives v' is non-zero (in each case, that for which $i = k$), and $\rho v' = \pi_m$. Thus for each integer k between 1 and m, $x - a_k$ is a factor of $\pi_m - \rho v'$, so v divides $\pi_m - \rho v'$.

The quotient, σ, is then another polynomial in $x, a_1, a_2, \ldots, a_m$ symmetric in all the a_i. Applying the Fundamental Theorem on Symmetric Polynomials to each term in the equation $\pi_m = \sigma v + \rho v'$ and replacing each elementary symmetric polynomial s_j therein by the indeterminate i_j produces the equation

$$(13) \qquad p_m = s(x) y(x) + r(x) y'(x),$$

where $r(x)$ and $s(x)$ are polynomials in $x, i_1, i_2, \ldots, i_m$, and $y(x)$ is the polynomial defined in (11). Replacing each i_j in (13) by the real number C_j then yields $P_Y = S(x)Y(x) + R(x)Y'(x)$, whose left member is a real number and whose right member is a polynomial in x. If $P_Y \neq 0$, division by P_Y yields $1 = \frac{S_Y(x)}{P_Y} Y(x) + \frac{R_Y(x)}{P_Y} Y'(x)$; so $Y(x)$ and $Y'(x)$ have no nonconstant common factor unless $P_Y = 0$.

Conversely, if $Y(x)$ and $Y'(x)$ have a nonconstant common factor, there are functions $f(x)$ and $\phi(x)$ such that $f(x)Y(x) + \phi(x)Y'(x) = 1$. Moving the left member of that equation to the right and adding $f(x)v + \phi(x)v'$ to both sides then gives

$$(14) \qquad f(x)v + \phi(x)v' = 1 + f(x)(v - Y(x)) + \phi(x)(v - Y(x))'.$$

Gauss abbreviates the expression $f(x)(y(x) - Y(x)) + \phi(x)(y(x) - Y(x))'$, a polynomial in the indeterminates $x, i_1, \ldots, i_m$, by $F(x, i_1, \ldots, i_m)$, and the right member of (14), regarded as a polynomial in x and the elementary symmetric polynomials $s_1, \ldots, s_m$, by $1 + F(x, s_1, \ldots, s_m)$. Similarly, he uses $F(x, C_1, \ldots, C_m)$ to denote the result of replacing each i_k in $F(x, i_1, \ldots, i_m)$ by C_k; and since $y(x)$ is thereby transformed into $Y(x)$, it follows that for any value of x,

$$(15) \qquad\qquad F(x, C_1, \ldots, C_m) = 0.$$

Gauss next applies the product rule to v, which, for any j between 1 and m, yields

$$(16) \qquad v' = \prod_{\substack{1 \le i \le m \\ i \ne j}} (x - a_i) + (x - a_j)\left(\prod_{\substack{1 \le i \le m \\ i \ne j}} (x - a_i) \right)'.$$

Replacing v' in (14) by the expression on the right of (16) and setting $x = a_j$ successively for each j between 1 and m then gives the sequence of equations

$$\phi(a_j) \prod_{\substack{1 \le i \le m \\ i \ne j}} (x - a_i) = 1 + F(a_j, s_1, \ldots, s_m) \qquad (1 \le j \le m).$$

Since the expressions on either side of each equation in that sequence are polynomials symmetric in the indeterminates $a_1, \ldots, a_m$, the same is true of the product of those equations:

$$(17) \qquad \pi_m \phi(a_1)\phi(a_2) \ldots \phi(a_m) = \prod_{i=1}^{m} (1 + F(a_i, s_1, \ldots, s_m)).$$

The Fundamental Theorem on Symmetric Polynomials thus ensures that there are polynomials in the indeterminates $i_1, \ldots, i_m$, say t and ψ, such that t is transformed into $\phi(a_1)\phi(a_2) \ldots \phi(a_m)$ and ψ into $\prod_{i=1}^{m}(1 + F(a_i, s_1, \ldots, s_m))$ when each i_j is replaced by s_j. From (17) it then follows that $p_m t = \psi$.

Replacing each indeterminate i_j in this last equation by the real number C_j, and writing T for the resulting value of t, we conclude from (15) that $P_Y T = 1$. Hence $P_Y \ne 0$.

Lemma 1, just proved, implies that if $P_Y = 0$, then $Y(x)$ must have a nontrivial factor; so repeating that argument, if need be, shows that it must in fact have a factor Q for which $P_Q \ne 0$. Hence without loss of generality we may assume that $P_Y \ne 0$. Moreover, if $Y(x)$ has degree $m = k2^\mu$, where k is odd, then at least one factor of $Y(x)$ must be of degree $l2^\tau$, where l is odd and $\tau \le \mu$, for otherwise the power of 2 in m would exceed μ.

Gauss went on in section 12 to consider the polynomial

$$(18) \qquad \zeta(u, x) = \prod [u - (a_i + a_j)x + a_i a_j],$$

where the product is taken over the $m(m - 1)/2$ unordered pairs $\{a_i, a_j\}, i \neq j$ of the indeterminates $a_1, \ldots, a_m$. Once again, ζ is symmetric in those indeterminates, so there is a function $z(u, x)$ in the indeterminates $i_1, \ldots, i_m$ that transforms to ζ when the latter indeterminates are replaced by $s_1, \ldots, s_m$, and a function $Z(u, x)$ that results from $z(u, x)$ when each s_i is replaced by C_i (the ith coefficient of Y). Regarded as functions of u alone, ζ and z are monic polynomials of degree $n = m(m - 1)/2$, with coefficients that are polynomials in $x, a_1, \ldots, a_m$ and in $x, i_1, \ldots, i_m$, respectively, and Z is a monic polynomial of degree n in u whose coefficients are polynomials in x, say $c_1(x), \ldots, c_n(x)$. We may then consider the discriminant $P_Z(x)$ of Z, that is, the function $p_n(c_1(x), \ldots, c_n(x))$. (See the last footnote above.)

Gauss's next task was to prove

Lemma 2: If $P_Y \neq 0$, then $P_Z(x)$ cannot be identically zero.

He noted that, once again, that would be straightforward if $Y(x)$ were a product of linear factors. To establish the result without that assumption, he observed that the discriminant of ζ is the product of all the $n(n - 1)$ non-zero differences of distinct pairs of the n expressions $(a_i + a_j)x - a_i a_j$. Hence the discriminants of ζ and of z, regarded as polynomials in x, each have degree $d = n(n - 1) = \frac{1}{4}m(m - 1)(m + 1)(m - 2)$, while the discriminant $P_Z(x)$ of Z may have lesser degree if the particular values of the C_i cause the coefficient of x^d in $P_Z(x)$ to vanish. The problem is to show that not all the coefficients of $P_Z(x)$ will vanish.

Closer examination of the discriminant of ζ reveals that that product may be split into two groups of factors, the first consisting of those differences of the form

$$[(a_i + a_j)x - a_i a_j] - [(a_i + a_k)x - a_i a_k] = (a_j - a_k)(x - a_i),$$

for distinct i, j, k, and the second of those differences of the form

$$(19) \quad [(a_i + a_j)x - a_i a_j] - [(a_k + a_l)x - a_k a_l] = (a_i + a_j - a_k - a_l)x - a_i a_j + a_k a_l,$$

for distinct i, j, k, l. In the first group, each factor $(a_j - a_k)$ will occur $m - 2$ times (once for each value of i distinct from j and k), whereas each factor $(x - a_i)$ will occur $(m - 1)(m - 2)$ times (once for every ordered pair of distinct values j, k different from i). If the product of the second group of factors (a polynomial symmetric in $a_1, \ldots, a_m$) be denoted by κ, then from (10) and (12), the discriminant of ζ is $\pi_m{}^{m-2} v^{(m-1)(m-2)} \kappa$.

Likewise, if $k(x, i_1, \ldots, i_m)$ is the function that transforms into κ when each i_j in it is replaced by s_j, and $K(x)$ is the result of replacing each such i_j by C_j, then the discriminant of z is $p_m{}^{m-2} y^{(m-1)(m-2)} k$ and that of Z, P_Z, is $P_Y{}^{m-2} Y^{(m-1)(m-2)} K$. Since $P_Y \neq 0$ by assumption, it remains to show that K is not identically zero.

Toward that end Gauss introduced the function of the new indeterminate w given by

(20)
$$\prod[(a_i + a_j - a_k - a_l)w + (a_i - a_k)(a_i - a_l)],$$

where i, j, k, l are distinct integers between 1 and m and no factors are repeated. (Note that each factor is invariant under the interchange of a_k and a_l, and so would appear twice in the product if repeated factors were allowed.) That product is symmetric in the indeterminates $a_1, \ldots, a_m$, so it is uniquely expressible as a polynomial function $f(w, s_1, \ldots, s_m)$ of the elementary symmetric polynomials and w. Since the number of factors in the product is $\frac{1}{2}m(m-1)(m-2)(m-3)$, the degree of each substitution instance of $f(w, s_1, \ldots, s_m)$ is at most that. Also,

$$f(0, s_1, \ldots, s_m) = \pi_m{}^{(m-2)(m-3)},$$

$$f(0, i_1, \ldots, i_m) = p_m{}^{(m-2)(m-3)}, \quad \text{and}$$

$$f(0, C_1, \ldots, C_m) = P_Y{}^{(m-2)(m-3)}.$$

In particular, the last equation shows that the constant term of

$$f(w, C_1, \ldots, C_m)$$

does not vanish.

Let the non-zero term of highest degree in $f(w, C_1, \ldots, C_m)$ be Nw^γ. Then for each j between 1 and m, if w be replaced by $x - a_j$, $f(x - a_j, C_1, \ldots, C_m)$ may be regarded as a polynomial in x whose leading term is Nx^γ and whose other coefficients depend upon a_j. Consequently

(21)
$$\prod_{j=1}^{m} f(x - a_j, C_1, \ldots, C_m)$$

is a polynomial in x with leading term $N^m x^{m\gamma}$, in which the coefficients of the remaining terms are functions of $a_1, \ldots, a_m$.

Similarly,

$$\prod_{j=1}^{m} f(x - a_j, i_1, \ldots, i_m)$$

is a polynomial function of $x, a_1, \ldots, a_m, i_1, \ldots, i_m$ symmetric in $a_1, \ldots, a_m$, which by the Fundamental Theorem on Symmetric Polynomials may be rewritten as a polynomial $\varphi(x, s_1, \ldots, s_m, i_1, \ldots, i_m)$. Replacing each i_j by s_j then yields

$$\varphi(x, s_1, \ldots, s_m, s_1, \ldots, s_m) = \prod_{j=1}^{m} f(x - a_j, s_1, \ldots, s_m).$$

Now, for any fixed value of i, when w is replaced by $x - a_i$, the factor

$$(a_i + a_j - a_k - a_l)w + (a_i - a_k)(a_i - a_l)$$

in (20) reduces after cancellation of like terms to

$$(a_i + a_j - a_k - a_l)x - a_i a_j + a_k a_l,$$

which is the same as the right member of (19). So each factor of κ is also a factor of $\varphi(x, s_1, \ldots, s_m, s_1, \ldots, s_m)$; that is, κ divides $\varphi(x, s_1, \ldots, s_m, s_1, \ldots, s_m)$, say $\varphi(x, s_1, \ldots, s_m, s_1, \ldots, s_m) = \kappa \chi(x, s_1, \ldots, s_m)$. Therefore also

$$\varphi(x, C_1, \ldots, C_m, C_1, \ldots, C_m) = K \chi(x, C_1, \ldots, C_m).$$

But $\varphi(x, s_1, \ldots, s_m, C_1, \ldots, C_m)$ is the product in (21), which has leading term $N^m x^{m\gamma}$, not involving any of $s_1, \ldots, s_m$; so $N^m x^{m\gamma}$ must also be the leading term of $\varphi(x, C_1, \ldots, C_m, C_1, \ldots, C_m)$. In particular, $\varphi(x, C_1, \ldots, C_m, C_1, \ldots, C_m)$ does not vanish identically. Therefore neither does K, as was to be shown.

Before beginning the induction that lay at the heart of his second proof, Gauss stated and proved one final lemma.

Lemma 3: Let $\Phi(u, x)$ denote the product $\prod_{i=1}^{k}(\alpha_i + \beta_i u + \gamma_i x)$ of any number of factors linear in the indeterminates u and x, and let v be another indeterminate. Then the function

$$\Omega = \Phi\left(u + v \frac{\partial \Phi}{\partial x}, x - v \frac{\partial \Phi}{\partial u}\right)$$

is divisible by $\Phi(u, x)$.

Proof: For each i between 1 and k, we have

$$\Phi(u, x) = (\alpha_i + \beta_i u + \gamma_i x) Q_i,$$

where Q_i denotes the product of all the factors $\alpha_j + \beta_j u + \gamma_j x$ with $j \neq i$. (Each Q_i is thus a polynomial in $u, x, \alpha_j, \beta_j, \gamma_j$, for $1 \leq j \leq k, j \neq i$.) So

$$\frac{\partial \Phi}{\partial x} = \gamma_i Q_i + (\alpha_i + \beta_i u + \gamma_i x)\frac{\partial Q_i}{\partial x} \quad \text{and} \quad \frac{\partial \Phi}{\partial u} = \beta_i Q_i + (\alpha_i + \beta_i u + \gamma_i x)\frac{\partial Q_i}{\partial u}.$$

Substituting the expressions on the right sides of those equations into the corresponding factor

$$\alpha_i + \beta_i u + \gamma_i x + \beta_i v \frac{\partial \Phi}{\partial x} - \gamma_i v \frac{\partial \Phi}{\partial u}$$

of Φ yields

$$(\alpha_i + \beta_i u + \gamma_i x)(1 + \beta_i v \frac{\partial Q}{\partial x} - \gamma_i v \frac{\partial Q}{\partial u}),$$

and consequently,

$$\Omega = \Phi(u, x) \prod_{i=1}^{k} (1 + \beta_i v \frac{\partial Q}{\partial x} - \gamma_i v \frac{\partial Q}{\partial u}). \qquad \text{q.e.d.}$$

When applied to $\zeta(u, x) = z(u, x, s_1, \ldots, s_m)$, Lemma 3 shows that ζ divides

$$z(u + v \frac{\partial \zeta}{\partial x}, x - v \frac{\partial \zeta}{\partial u}, s_1, \ldots, s_m),$$

say with quotient $\Psi(u, x, v, s_1, \ldots, s_m)$. Likewise,

$$z(u + v \frac{\partial z}{\partial x}, x - v \frac{\partial z}{\partial u}, i_1, \ldots, i_m) = z(u, x) \Psi(u, x, v, i_1, \ldots, i_m)$$

and

$$Z(u + v \frac{\partial Z}{\partial x}, x - v \frac{\partial Z}{\partial u}, C_1, \ldots, C_m) = Z(u, x) \Psi(u, x, v, C_1, \ldots, C_m).$$

Now, given definite values U and X for the indeterminates u and x, let U' and X' denote

$$\frac{\partial Z}{\partial u} \Big|_{(U,X)} \quad \text{and} \quad \frac{\partial Z}{\partial x} \Big|_{(U,X)},$$

respectively. Then

$$Z(U + vX', X - vU') = Z(U, X) \Psi(U, X, v, C_1, \ldots, C_m).$$

If $U' \neq 0$, replacing v by $\dfrac{X - x}{U'}$ yields

$$(22) \qquad Z(U + \frac{XX'}{U'} - \frac{xX'}{U'}, x) = Z(U, X) \Psi(U, X, \frac{X - x}{U'}, C_1, \ldots, C_m).$$

In other words, setting $u = U + \dfrac{X - x}{U'} X'$ transforms $Z(u, X)$ into

$$Z(U, X) \Psi(U, X, \frac{X - x}{U'}, C_1, \ldots, C_m).$$

By Lemma 2, the assumption that $P_Y \neq 0$ implies that the polynomial P_Z does not identically vanish. Therefore $P_Z(x) = 0$ for only finitely many values of x, so a real number X may be chosen for which $P_Z(X) \neq 0$; that is, the discriminant of the function $Z(u, X)$ is non-zero. By Lemma 1, the polynomial $Z(u, X)$ and its derivative $\dfrac{dZ}{du}$ thus have no common factor. Also, as noted earlier, $Z(u, X)$ has degree $n = m(m - 1)/2$ in u, where $m = k2^\mu$, k odd, is the degree of $Y(x)$. Hence $n = (k2^\mu)(k2^\mu - 1)/2 = (k^2)2^{2\mu-1} - k2^{\mu-1} = [k(k2^\mu - 1)]2^{\mu-1}$. The quantity in brackets in the last member of that equation is odd, so the power of 2 in n is less than the power of 2 in m. Therefore we may assume by induction that there is a real or complex value U for which $Z(U, X) = 0$.[16] By the factor theorem, $u - U$ must then be a factor of $Z(u, X)$, but, *ipso facto*, not a factor of $\dfrac{dZ}{du}$. So $U' \neq 0$, again by the factor theorem.

For those particular values of X and U, the right-hand member of (22) is identically zero, independent of the value of x. Thus $Z(u, x)$, regarded as a polynomial in u with coefficients that are polynomials in x, vanishes when $u = U + \dfrac{X - x}{U'}X'$, and so has $u - U - \dfrac{X - x}{U'}X'$ as a factor. If we then let $u = x^2$, the polynomial $Z(x^2, x)$ must have the quadratic polynomial

$$x^2 - U - \frac{X - x}{U'}X' = x^2 + \frac{X'}{U'}x - (U + \frac{XX'}{U'})$$

as a factor. The quadratic formula then provides a real or complex root of $Z(x^2, x)$.

Finally, recall that $z(x^2, x, i_1, \ldots, i_m)$ is the unique polynomial that transforms into $\zeta(x^2, x)$ when each i_j is replaced by the elementary symmetric polynomial s_j, and that $Z(x^2, x)$ is obtained from $z(x^2, x, i_1, \ldots, i_m)$ by replacing each i_j by the coefficient C_j of $Y(x)$. But

$$\zeta(x^2, x) = \prod [x^2 - (a_i + a_j)x + a_i a_j] = \prod (x - a_i)(x - a_j)$$

$$= \prod_{i=1}^{m} (x - a_i)^{m-1} = v^{m-1}$$

(where the first two products are taken over all unordered pairs $\{a_i, a_j\}, i \neq j$; cf. (18) and (12) above), and the unique polynomial in $x, i_1, \ldots, i_m$ that transforms into v^{m-1} when each i_j is replaced by s_j is $y(x)^{m-1}$ (cf. (11)). Therefore $z(x^2, x) = (y(x))^{m-1}$ and $Z(x^2, x) = (Y(x))^{m-1}$, so the root found for $Z(x^2, x)$ is also a root of $Y(x)$, completing the proof of the theorem.

---

[16]Gauss observed that the coefficients of $Z(u, X)$ will be real numbers if X is real and all coefficients of $Y(x)$ are real — a fact needed for the base case of the induction (that a real polynomial of odd degree must have a real root).

The proof just given is remarkable not only for purity of method but for its economy of means. The principal tool invoked is the Fundamental Theorem on Symmetric Polynomials, applied over and over again to a sequence of carefully chosen polynomials, and the only non-algebraic principle used is the intermediate-value theorem. As such it is a *tour de force*. Its length, however, is a pedagogical deterrent, and the justifications for some of the definitions and substitutions employed become apparent only with hindsight; it is thus less perspicuous than Argand's nearly contemporaneous argument.

8.6   Proofs based on integration

The proofs of the Fundamental Theorem of Algebra discussed above are all direct proofs. There are indirect proofs as well, several of which are based on the theory of integration. Among them is another by Gauss, published just one year after his second.[17]

Gauss's third proof: As in his first proof, given a monic polynomial

$$Y = x^m + A_1 x^{m-1} + \cdots + A_{m-1} x + A_m$$

with real coefficients, Gauss began by writing the variable x in polar form as $x = r(\cos\phi + i \sin\phi)$ and considered the real and imaginary parts of Y, which he denoted by t and u. Thus (replacing Gauss's $A, B, \ldots, M$ by $A_1, \ldots, A_m$)

$$t = \sum_{j=0}^{m} A_j r^{m-j} \cos(m-j)\phi \quad \text{and} \quad u = \sum_{j=0}^{m} A_j r^{m-j} \sin(m-j)\phi,$$

where $A_0 = 1$ and, for $1 \le j \le m$, the coefficients A_j are arbitrary real numbers. He further defined

$$t' = \sum_{j=0}^{m} (m-j) A_j r^{m-j} \cos(m-j)\phi,$$

$$u' = \sum_{j=0}^{m} (m-j) A_j r^{m-j} \sin(m-j)\phi,$$

$$t'' = \sum_{j=0}^{m} (m-j)^2 A_j r^{m-j} \cos(m-j)\phi,$$

---

[17]Translated into German on pp. 61–67 of Gauss (1890).

$$u'' = \sum_{j=0}^{m} (m-j)^2 A_j r^{m-j} \sin(m-j)\phi, \quad \text{and}$$

$$y = \frac{(t^2+u^2)(tt''+uu'') + (tu'-ut')^2 - (tt'+uu')^2}{r(t^2+u^2)^2},$$

and stipulated that R should be a real number greater than the largest of the numbers $(m|A_j|\sqrt{2})^{1/j}$, for $1 \le j \le m$. He then claimed that setting $r = R$ would ensure that $tt' + uu'$ was positive, for any angle ϕ.

Proof of claim: Corresponding to the definitions of t, u, t' and u', let

$$T = \sum_{j=0}^{m} A_j R^{m-j} \cos(\frac{\pi}{4} + j\phi),$$

$$U = \sum_{j=0}^{m} A_j R^{m-j} \sin(\frac{\pi}{4} + j\phi),$$

$$T' = \sum_{j=0}^{m} (m-j) A_j R^{m-j} \cos(\frac{\pi}{4} + j\phi), \quad \text{and}$$

$$U' = \sum_{j=0}^{m} (m-j) A_j R^{m-j} \sin(\frac{\pi}{4} + j\phi).$$

Gauss observed that T could be rewritten as

$$\sum_{j=1}^{m} \frac{R^{m-j}}{m\sqrt{2}} (R^j + mA_j \sqrt{2} \cos(\frac{\pi}{4} + j\phi)),$$

and similarly for U, T' and U'; so, by the stipulation on R, those four quantities, and hence $TT' + UU'$ must all be positive. But when $r = R$, $tt' + uu' = TT' + UU'$. To see that, note first that when $r = R$, the quantity t is equal to

$$(23) \qquad\qquad T \cos(\frac{\pi}{4} + m\phi) + U \sin(\frac{\pi}{4} + m\phi).$$

For, by the definitions of T and U, each term of (23) is of the form

$$(24) \quad A_j R^{m-j} [\cos(\frac{\pi}{4} + j\phi) \cos(\frac{\pi}{4} + m\phi) + \sin(\frac{\pi}{4} + j\phi) \sin(\frac{\pi}{4} + m\phi)]$$

$$= \frac{A_j R^{m-j}}{2} [\cos((m-j)\phi) + \cos(\frac{\pi}{2} + (m+j)\phi) + \cos((m-j)\phi)$$

$$- \cos(\frac{\pi}{2} + (m+j)\phi)] = A_j R^{m-j} \cos((m-j)\phi),$$

the corresponding term of t. Likewise, when $r = R$, the quantities u, t' and u' are equal to

$$T \sin(\frac{\pi}{4} + m\phi) - U \cos(\frac{\pi}{4} + m\phi),$$

$$T' \cos(\frac{\pi}{4} + m\phi) + U' \sin(\frac{\pi}{4} + m\phi), \quad \text{and}$$

$$T' \sin(\frac{\pi}{4} + m\phi) - U' \cos(\frac{\pi}{4} + m\phi),$$

respectively. Then, letting $A = \dfrac{\pi}{4} + m\phi$, we have

$$tt' = TT'\cos^2 A + T'U \sin A \cos A + TU' \sin A \cos A + UU'\sin^2 A \quad \text{and}$$

$$uu' = TT'\sin^2 A - T'U \sin A \cos A - TU' \sin A \cos A + UU'\cos^2 A,$$

so $tt' + uu' = TT' + UU' > 0$ when $r = R$, as claimed.

In addition, when $r = R$,

$$t^2 = T^2\cos^2 A + 2TU \sin A \cos A + U^2\sin^2 A \quad \text{and}$$

$$u^2 = T^2\sin^2 A - 2TU \sin A \cos A + U^2\cos^2 A,$$

so $t^2 + u^2 = T^2 + U^2$. Consequently, for any r satisfying the stipulations on R, $t^2 + u^2$ must be positive, whence t and u cannot simultaneously equal 0.

On the other hand, within the circle C of radius R centered at the origin there must be a point (r, ϕ) where both $t = 0$ and $u = 0$ (and thus a point $x = r(\cos\phi + i \sin\phi)$ where $Y = 0$, proving the theorem). For suppose not. Then let

$$\Omega = \int\int_C y \, dA = \int_0^R \int_0^{2\pi} y \, d\phi \, dr = \int_0^{2\pi} \int_0^R y \, dr \, d\phi.$$

Note that $\dfrac{\partial t}{\partial \phi} = -u', \dfrac{\partial u}{\partial \phi} = t', \dfrac{\partial t'}{\partial \phi} = -u''$ and $\dfrac{\partial u'}{\partial \phi} = t''$. Using those relations, one computes that

$$(25) \qquad \frac{\partial}{\partial \phi}\left[\frac{tu' - ut'}{r(t^2 + u^2)}\right] = y, \quad \text{that is,} \quad \int y \, d\phi = \frac{tu' - ut'}{r(t^2 + u^2)}.$$

Since u and u' both equal 0 when $\phi = 0$ or $\phi = 2\pi$, the last expression above is also zero for those values of ϕ, whence

$$(26) \qquad \int_0^{2\pi} y \, d\phi = 0, \quad \text{and so} \quad \Omega = \int_0^R \int_0^{2\pi} y \, d\phi \, dr = 0.$$

Similarly, from $r\dfrac{\partial t}{\partial r} = t'$, $r\dfrac{\partial u}{\partial r} = u'$, $r\dfrac{\partial t'}{\partial r} = t''$, and $r\dfrac{\partial u'}{\partial r} = u''$, one computes that

(27) $\qquad \dfrac{\partial}{\partial r}\left[\dfrac{tt' + uu'}{t^2 + u^2}\right] = y,\quad$ that is, $\quad\displaystyle\int y\, dr = \dfrac{tt' + uu'}{t^2 + u^2}.$

Consequently,

$$\int_0^R y\, dr = \frac{tt' + uu'}{t^2 + u^2}\Big|_0^R = \frac{TT' + UU'}{T^2 + U^2} > 0 \quad\text{by the claim proved earlier.}$$

But then

$$\Omega = \int_0^{2\pi}\int_0^R y\, dr\, d\phi = \int_0^{2\pi}\frac{TT' + UU'}{T^2 + U^2} > 0,\quad\text{contrary to (26).}$$

In his prefatory remarks, Gauss said merely that continued reflection on the Fundamental Theorem had led him to this third proof, which, like the second, was "purely analytic," but was based on entirely different principles and far surpassed the second in simplicity. And indeed, like Argand's proof, nothing beyond advanced calculus is needed for understanding the argument just given. However, several of the functions used in the proof, especially y, are introduced seemingly out of the blue, and it seems almost miraculous that the partial derivatives in (25) and (27) turn out to equal y. Thus, though succinct and requiring minimal prerequisites, the proof is not perspicuous: It provides convincing *verification* that the Fundamental Theorem is true, but it is not explanatory, since it does not convey understanding of *why* it is.

Other indirect proofs of the Fundamental Theorem are based on Cauchy's theory of complex contour integration. The best known is perhaps that based on **Liouville's Theorem** (that a bounded entire function must be constant): For if the polynomial $p(z)$ had no zero in the complex plane, then $\dfrac{1}{p(z)}$ would be an entire function; and as in Argand's proof, for any positive constant C there is an $R > 0$ such that $|p(z)| > C$ whenever $|z| > R$. Thus $\dfrac{1}{p(z)} < \dfrac{1}{C}$ for $|z| > R$, and as a continuous function, $\dfrac{1}{p(z)}$ would also be bounded within the disc $|z| \le R$. Thus $\dfrac{1}{p(z)}$ would be bounded throughout the complex plane, and hence a constant by Liouville's Theorem — a contradiction for any $p(z)$ of positive degree.

Liouville's Theorem is itself a consequence of **Cauchy's integral formula**, which asserts that if $f(z)$ is any function analytic in a simply connected domain containing the simple closed curve γ, then for any point z_0 inside γ,

$$f(z_0) = \frac{1}{2\pi i} \int_\gamma \frac{f(z)}{z - z_0}\, dz;$$

and, as first noted in Zalcman (1978) (see also Lax and Zalcman 2012), Cauchy's formula may be applied directly to yield an even simpler proof of the Fundamental Theorem of Algebra. For if $|p(z)| \neq 0$ throughout the complex plane, then $q(z) = \frac{1}{p(z)}$ is entire and $\frac{1}{p(0)} = q(0) \neq 0$. Hence

$$q(0) = \frac{1}{2\pi i} \int_{|z|=R} \frac{q(z)}{z}\, dz = \frac{1}{2\pi} \int_0^{2\pi} q(Re^{i\theta})\, d\theta,$$

for any $R > 0$. But as R approaches ∞, the last integral approaches zero, contrary to $q(0) \neq 0$.

Alternatively, a proof of the Fundamental Theorem may be couched in terms of **winding numbers**, where the winding number of a continuously differentiable closed curve γ about the origin is given by

$$\frac{1}{2\pi i} \int_\gamma \frac{dz}{z},$$

if γ does not pass through the origin. That notion can, however, also be defined without reference to line integrals: Less formally, and more generally, if $f(z)$ is a continuous function that is never zero, the winding number of $f(z)$ around the origin as z traces out a continuously differentiable closed curve γ may be defined, as in Courant and Robbins (1941), as "the net number of complete revolutions made by a vector joining the origin to $f(\gamma(z))$ as z traces out γ." Courant and Robbins then offer the following indirect proof of the Fundamental Theorem.

Suppose that the monic polynomial $p(z) = z^n + a_{n-1}z^{n-1} + \cdots + a_0$ of degree $n > 0$ is never zero. Let C_t be the circle about the origin of radius t, given by the equation $z = te^{i\theta}$, and let $\phi(t)$ be the winding number of $p(z)$ around the origin as z traces out C_t. Then ϕ is a continuous, integer-valued function of t, and so must be a *constant*; and since $\phi(0) = 0$, we must have $\phi(t) = 0$ for all t.

But

$$|z^n - p(z)| \leq |a_{n-1}||z|^{n-1} + \cdots + |a_0| = |z|^{n-1}\left[|a_{n-1}| + \cdots + \frac{|a_0|}{|z|^{n-1}}\right],$$

so for values of t greater than $|a_0| + |a_1| + \cdots + |a_{n-1}| + 1$, the length $|z^n - p(z)|$ of the vector from $p(z)$ to z^n will be less than or equal to

$$t^{n-1}[|a_{n-1}| + \cdots + \frac{|a_0|}{t^{n-1}}] < t^n = |z^n|, \text{ the distance from } z^n \text{ to the origin.}$$

(See Figure 8.4.)

Fig. 8.4 A winding-number
proof. Adapted from Figure
150, p. 270 in *What is*
Mathematics?, 2nd. ed.
(1996), by Richard Courant
and Herbert Robbins. By
permission of Oxford
University Press, U.S.A.

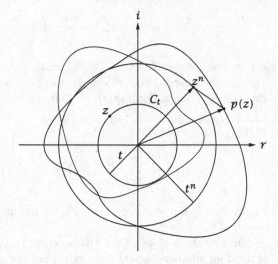

The segment joining $p(z)$ to z^n thus cannot pass through the origin when z is
on C_t. Deforming the curve traced by $p(z)$ to the circle traced by z^n by shrinking
each such line segment to zero will thus not alter the value of $\phi(t)$, which must be
the same as the winding number of z^n around the origin as z traces out C_t.[18] But that
number is n, which is greater than zero, contrary to what was found above.

The preface to the first edition of *What is Mathematics?* states that that book
"presupposes only knowledge that a good high school course could impart," and
the proof just given is an excellent example of a perspicuous informal proof. By
dispensing with reference to line integrals, Courant and Robbins succeeded in
offering a proof of the Fundamental Theorem that should be both convincing and
understandable to mathematically inclined high school students — a remarkable
achievement from a pedagogical standpoint, since it is at that level that students
first encounter the Fundamental Theorem, and all other proofs known to this author
presume at least knowledge of advanced calculus.[19]

---

[18] The formal statement of that fact is Rouché's Theorem.

[19] I vividly recall my own frustration on repeatedly reading, not only in high school, but throughout
my undergraduate courses at M.I.T., that a proof of the Fundamental Theorem was "beyond
the scope of this text." Only when I took a graduate complex analysis course did I finally see
the theorem proved as a corollary to Liouville's theorem — an experience I found distinctly
anticlimactic, given my long years of expectant waiting. (William Dunham, in his article Dunham
(1991), describes his own very similar experience.)

8.7 Other modern proofs

As part of the development of field theory in the nineteenth century, Kronecker proved the basic result needed to establish the existence of splitting fields: that if $p(x)$ is an irreducible polynomial with coefficients in a field F, then there is a field F' extending F, of finite degree over F, in which $p(x)$ has a root. The earlier arguments put forward as proofs of the Fundamental Theorem of Algebra by Lagrange and Laplace were thereby validated.

Later in the nineteenth century the work of Évariste Galois was belatedly published, and in 1872 Peter Ludvig Sylow proved the theorem in group theory now bearing his name, according to which for any prime number p, if p^m is the largest power of p dividing the order of a finite group G, then G must possess subgroups of order p^k for each $k \leq m$. In addition, the number of subgroups of G of order p^m divides the order of G and is congruent to 1 modulo p. Emil Artin then gave a proof of the Fundamental Theorem in terms of those concepts.

Proof via Galois theory: [20] Let K be a splitting field for the polynomial $p(x)$, of degree $n > 1$ with real coefficients. K is a finite extension of the real field $\mathbb{R}$, say of degree $d = 2^m q$ over $\mathbb{R}$, where q is odd. Since $\mathbb{R}$ has characteristic 0, K is a simple extension $\mathbb{R}(\alpha)$ of $\mathbb{R}$, where α is a root of a unique monic irreducible polynomial $r(x)$ of degree d with real coefficients. If $m = 0$, $r(x)$ must have degree $q = 1$, since every real polynomial of odd degree has a real root. So in that case $K = \mathbb{R}$. If $m = 1$ and $q = 0$, $r(x)$ is a quadratic polynomial, whose roots must lie in $\mathbb{C}$, whence $K = \mathbb{C}$. So suppose $m > 0$ and $q \neq 0$, and let G be the Galois group of K over R. By the fundamental theorem of Galois theory, G must have order d, and by Sylow's theorem, G therefore has a subgroup of order 2^m and index q. Again by the fundamental theorem of Galois theory, there must then be an extension field E of $\mathbb{R}$ intermediate between $\mathbb{R}$ and K, having degree q over $\mathbb{R}$. Then as above, since q is odd, $q = 1$ and $E = \mathbb{R}$. Accordingly, K must have degree 2^m over $\mathbb{R}$, so the order of G is 2^m. By Sylow's Theorem, G has a subgroup H_1 of order 2^{m-1}. Let L_1 be the fixed field of H_1. Then L_1 is an extension of $\mathbb{R}$ of degree 2, so L_1 is isomorphic to $\mathbb{C}$ (since L_1 is obtained from $\mathbb{R}$ by adjoining a root of a quadratic polynomial), and K is an extension of L_1 of degree 2^{m-1}. If $m > 1$, let H_2 be a subgroup of H_1 of order 2^{m-2}, and let L_2 be the fixed field of H_2. Then L_2 is an extension of L_1 (and so is isomorphic to an extension of $\mathbb{C}$) of degree 2. But that is impossible, since quadratic polynomials with complex coefficients always have complex roots. Thus $m = 1$ and K has degree $2^0 = 1$ over L_1; that is, $K \simeq L_1 \simeq \mathbb{C}$, where $\simeq$ denotes isomorphism.   q.e.d.

The prerequisites for understanding the proof just given are substantially greater than those for the other proofs so far considered. In particular, few undergraduates would be able to comprehend it. Nonetheless, since Galois theory was developed to

---

[20]For further background details, see chapter 7 of Fine and Rosenberger (1997).

answer questions about the solvability of polynomials by radicals, it is appropriate to use it to prove the Fundamental Theorem of Algebra; and like Gauss's second proof, it is as methodologically pure as possible, invoking nothing non-algebraic except the intermediate-value theorem.

There are also proofs of the Fundamental Theorem based on sophisticated notions from algebraic topology (Brouwer degree, homotopy theory, simplicial complexes, homology theory), three of which are sketched in chapter 9 and appendix D of Fine and Rosenberger (1997). They are not discussed further here, since the means they employ were developed to address very different concerns and go far beyond what is required to establish the Fundamental Theorem. Nevertheless, they demonstrate the power of topological techniques (another example of bench-marking) and illustrate the coherence of disparate mathematical theories.

A much simpler proof, based on notions from point-set topology, establishes the Fundamental Theorem in the equivalent form: Every polynomial of non-zero degree with complex coefficients, regarded as a mapping with domain $\mathbb{C}$, has range all of $\mathbb{C}$.[21]

A topological proof:[22] Let $p(z)$ be a polynomial of degree $n > 1$ with complex coefficients. Since p is continuous and $|p(z)|$ approaches infinity as $|z|$ does, the range R of p must be closed (by Weierstrass's theorem that every bounded sequence has a convergent subsequence). Consider then the set T of all points $p(z)$ where $p'(z) = 0$. Since $\mathbb{C} - R$ is open, it suffices to show that $R - T$ is open as well. For if so, then $(\mathbb{C} - R) \cup (R - T) = \mathbb{C} - T$. Since T is finite, $\mathbb{C} - T$ is a connected set, which cannot be the union of two disjoint nonempty open sets; and since $p^{-1}(T)$ is finite and $n > 1$, $R - T$ is nonempty. Thus $\mathbb{C} - R$ must be empty.

The proof is completed by noting that for every $p(z_0)$ in $R - T$, $p'(z_0) \neq 0$, so the inverse function theorem implies that there are neighborhoods U of z_0 and V of $p(z_0)$ such that p maps U one-to-one onto V. Hence every point of $R - T$ is an interior point.

There remains the purity question broached earlier in connection with Euler's failed proof attempt: Can the original statement of the Fundamental Theorem, that any non-constant polynomial with real coefficients can be expressed as a product of real linear and quadratic factors, be proved without reference to complex numbers or splitting fields?

Such a proof has been given, based on concepts from linear algebra. The proof is by induction on the degree of the polynomial and uses properties of the so-called Bezoutian resultant (the determinant of an $n \times n$ symmetric matrix defined for any pair of polynomials p, q, where n is the maximum of the degrees of p and q). Specifically, writing p as a function of the variable x and q as a function of the variable w, the factor theorem implies that $x - w$ must be a factor of the

---

[21] The intermediate-value theorem may similarly be cast as the assertion that the range of any polynomial of *odd* degree with *real* coefficients, regarded as a mapping from $\mathbb{R}$ into $\mathbb{R}$, is all of $\mathbb{R}$.

[22] This is essentially the proof given in Sen (2000).

polynomial $P(x, w) = p(x)q(w) - p(w)q(x)$, say $P(x, w) = (x - w)b(x, w)$, where b is a polynomial each of whose terms is of the form $b_{ij}x^{i-1}w^{j-1}$, for $1 \leq i, j \leq n$. The coefficients b_{ij} are the entries of the Bezoutian matrix $B(p, q)$, which is nonsingular if and only if p and q have no common root.[23]

In outline, the proof then proceeds as follows: Given a real polynomial $p(x) = a_n x^n + a_{n-1}x^{n-1} + \cdots + a_0$ with $a_0 \neq 0$ and $a_n \neq 0$, if n is odd, then $p(x)$ has a real root x_0 by the intermediate-value theorem, so $p(x) = (x - x_0)q(x)$, where $q(x)$ has degree $n - 1$. By the induction hypothesis, $q(x)$ then is a product of real linear and quadratic factors, whence so is $p(x)$. If n is even, say $n = 2m$, then let r be a real parameter and define auxiliary polynomials $p_1(x)$ and $p_2(x)$ by

$$p_1(x) = p(rx) = a_n r^n x + a_{n-1} r^{n-1} x^{n-1} = \cdots + a_0 \quad \text{and}$$

$$p_2(x) = a_0 x^n + a_1 r x^{n-1} + \cdots + a_n r^n .$$

The intermediate-value theorem, together with properties of principal minors, can then be invoked to show that there must be a value r_0 of r for which the determinant of $B(p_1, p_2) = 0$; so when $r = r_0$, $p_1(x)$ and $p_2(x)$ will have some maximal common factor $f(x)$. Writing $p_1(x) = f(x)Q_1(x)$ and $p_2(x) = f(x)Q_2(x)$ yields $p_1(x)Q_2(x) = p_2(x)Q_1(x)$, where Q_1 and Q_2 are relatively prime polynomials of degree less than n. By the induction hypothesis, if Q_2 is not constant, it must be a product of linear and quadratic factors, all of which divide p_1. The quotient $p_1(x)/Q_2(x)$ then also has degree less than n, so it too must be a product of linear and quadratic factors. Hence $p_1(x)$ is a product of linear and quadratic factors, and replacing x in each of those factors by x/r_0 produces a similar factorization of $p(x)$.

On the other hand, if $Q_2(x)$ is a constant, then $f(x)$ has degree n, so $Q_1(x)$ is also a constant. Hence $p_1(x) = cp_2(x)$ for some constant c, so for each i between 0 and n, the coefficient of x^i in $p_1(x)$ must equal c times the coefficient of x^i in $p_2(x)$; that is, $a_i r_0^i = ca_{n-i} r_0^{n-i}$ for each such i. In particular, for $i = m$, $a_m r_0^m = ca_m r_0^m$, so $c = 1$ and $a_j r_0^j = a_{n-j} r_0^{n-j}$ for $0 \leq j \leq n$, which means that $p_1(x)$ is a palindromic (self-reciprocal) polynomial. But then either $x + 1$ is a factor of $p_1(x)$, or $p_1(x) = x^m q(X)$, where q is a polynomial of degree m and $X = x + x^{-1}$.

By the induction hypothesis, q is a product of real linear and quadratic polynomials in X, so $p_1(x)$ is a product of real linear, quadratic, and quartic polynomials in x. Since quartics can always be factored (e.g., by Euler's technique) into products of real linear and quadratic expressions, so can $p_1(x)$. The desired factorization of $p(x)$ is then obtained once again by replacing x everywhere by x/r_0.

The proof just given (published in Eaton 1960) employs very different techniques than the others considered here. However, the idea of using the resultant of two polynomials as a tool in proving the Fundamental Theorem of Algebra is not new.

---

[23]Indeed, the nullity of $B(p, q)$ equals the number of common zeroes of p and q, counting multiplicities.

Indeed, the British mathematician James Wood gave an incomplete proof of that sort in 1798, the year before Gauss's first proof.[24] A full proof using resultants was later published by Paul Gordan, who presented it as a simpler alternative to Gauss's second proof (Gordan 1876).

8.8   Constructive proofs

Although the foregoing proofs establish the *existence* of a root for any nonconstant polynomial, they provide no means for actually *finding* one. One may therefore wonder, as Weierstrass did in 1891, whether it is possible "for any given polynomial f in $\mathbb{C}(Z)$, to produce a sequence z_n of complex numbers by an effectively defined procedure, so that $|f(z_n)|$ is sufficiently small in relation to $|f(z_{n-1})|$ that it converges to a zero of f?"[25] Beyond that, one might ask whether the proof that such a procedure converges can be done constructively, and how computationally *efficient* the procedure is. The importance of such questions to mathematical practice is indicated by the appearance in 1969 of a volume entitled *Constructive Aspects of the Fundamental Theorem of Algebra* (Dejon and Henrici 1969), containing the proceedings of a symposium on that topic held two years before at the IBM Research Laboratory in Zürich.

In 1924 Hermann Weyl gave an intuitionistic proof of the Fundamental Theorem of Algebra that invoked winding numbers (Weyl 1924, pp. 142–146). Later Hellmuth Kneser presented a modification of Argand's proof in which, given a nonconstant polynomial $p(x)$, he defined a sequence of complex numbers and proved, also by means acceptable to intuitionists, that it converged to a root of p (H. Kneser 1940). Specifically, given a monic polynomial $p(x) = x^n + a_{n-1}x^{n-1} + \cdots + a_0$ of degree $n > 1$ with complex coefficients, Kneser defined a sequence of complex numbers x_m designed to make the ratio $|p(x_{m+1})|/|p(x_m)|$ as small as possible. Toward that end he expressed $p(x_m + y)$ as $p(x_m) + \sum_{i=1}^{n} b_i y^i$, chose y so that one of the terms in the sum would have the same argument as $-p(x_m)$ and would strongly dominate the other terms, and set $x_{m+1} = x_m + y$. To find such a y he employed a lemma whose proof was lengthy and delicate. Forty-one years later his son Martin published a simplified version of the proof (M. Kneser 1981) based on the following much simpler lemma:

Given a monic polynomial $p(x)$ of degree $n > 1$ with constant term a_0, there is a positive number $q < 1$, depending only on n, such that if $|a_0| \leq c$, a complex number y can be specified for which $|y| \leq c^{1/n}$ and $|p(y)| \leq qc$.

---

[24]Wood (1798), his only published mathematical paper. An analysis of Wood's argument, as well as a completion of it, has recently been given by Frank Smithies (Smithies 2000). I am indebted to Peter Goddard for bringing both those papers to my attention.

[25]Quoted on p. 115 of Remmert (1990) from the third volume of Weierstrass's *Mathematische Werke*, pp. 251–269.

The sequence $\{x_m\}$ may then be defined inductively, as follows, starting with $x_0 = 0$. Suppose that x_m has already been defined and satisfies $|p(x_m)| \leq q^m c$. Apply the lemma to $f(x) = p(x_m + x)$, which has constant term $p(x_m)$, and set $x_{m+1} = x_m + y$. Then $|x_{m+1} - x_m| = |y| \leq (q_m c)^{1/n}$ and $|p(x_{m+1})| \leq q(q^m c) = q^{m+1} c$. The first of those inequalities shows that the sequence $\{x_m\}$ converges to some value x and the second that the inductive hypothesis is satisfied by x_{m+1}, so that $p(x_m)$ converges to 0. Since p is continuous, it follows that $p(x) = 0$. The proof of the lemma is constructive, and can be made to satisfy intuitionistic demands as well.

Two years before Martin Kneser's paper appeared, Steve Smale and Morris Hirsch also defined a sequence of values guaranteed to converge to a root of any given polynomial of degree $n > 0$ (Hirsch and Smale 1979, pp. 303–309). Their procedure was based on a modification of Newton's method, and again involved an auxiliary proposition, namely:

For any positive integer n there exist real numbers $\sigma_1, \ldots, \sigma_n$ with $0 < \sigma_k \leq 1$ for $k = 1, \ldots, n$ and K_n satisfying $0 < K_n < 1$, such that if $h_k = \sigma_k e^{i\pi/k}$, then for any n-tuple $(v_1, \ldots, v_n)$ in $\mathbb{C}^n - 0$ there is an m for which

$$\left| 1 + \sum_{k=1}^{n} \left(\frac{v_k}{v_m} h_m \right)^k \right| < K_n.$$

Suppose then that $p(x)$ is a polynomial of degree $n > 0$ and z is any complex number. If $p(z) = 0$, let $z' = z$. Otherwise, for each positive integer $k \leq n$, let v_k be a kth root of $p^{(k)}(z)/k!p(z)$. Since $n > 0$, $p^{(n)}(z) \neq 0$, so $(v_1, \ldots, v_n)$ is in $\mathbb{C}^n - 0$. The auxiliary proposition then yields a $v_m \neq 0$, whence z' can be taken to be $z + h_m/v_m$. Then by Taylor's theorem,

$$|p(z')| = \left| p(z) + \sum_{k=1}^{n} \frac{p^{(k)}(z)}{k!} \left(\frac{h_m}{v_m} \right)^k \right|$$

$$= \left| p(z) \left[1 + \sum_{k=1}^{n} \frac{p^{(k)}(z)}{p(z)k!} \left(\frac{h_m}{v_m} \right)^k \right] \right|$$

$$= |p(z)| \left| 1 + \sum_{k=1}^{n} \left(\frac{v_k}{v_m} h_m \right)^k \right| < K_n |p(z)|.$$

To prove the Fundamental Theorem of Algebra, let z_0 be any complex number, and assume by induction that $z_0, \ldots, z_j$ have already been defined. If $p(z_j) = 0$, then z_j is a root of p, so put $z_{j+1} = z_j$. Otherwise define z_{j+1} to be $z_j + h_{m_0}/v_{m_0}$, where m_0 is the least value of m that satisfies the inequality given in the auxiliary proposition. Then the equations and inequality displayed above show that $|p(z_{j+1})| < K_n |p(z_j)|$. So $|p(z_j)| < (K_n)^j |p(z_0)|$ for every j. Since $K_n < 1$, $p(z_j)$ therefore converges to 0 as j approaches ∞. The convergence may be very slow, however, since the authors note that for large values of n, K_n is very close to 1.[26]

That a subsequence of $\{z_j\}$ must converge to a root of p follows by a compactness argument:[27] For since $|p(z)|$ approaches ∞ as z does, and since $|p(z_j)| < K_n |p(z_0)|$ for all j, all the values z_j must lie within some disc $|z| \leq R$. If there are infinitely many distinct values z_j, a subsequence of them must approach some point within that disc, since otherwise for each point z with $|z| \leq R$ some disc D_z centered at z would contain at most one of the z_j (namely z_j if and only if $z = z_j$). But since the disc $|z| \leq R$ is compact, a finite number of those D_z would cover it. Thus either the sequence $\{z_j\}$ is eventually constant, say $z_k = z_j$ for all $k \geq j$ (which by construction implies that $p(z_j) = 0$), or else every neighborhood of some z^* in $\mathbb{C}$ contains infinitely many of the z_j. In the latter case, the continuity of p implies that $p(z^*) = 0$, so z^* is a root of p.

Further algorithms and constructive proofs of the Fundamental Theorem of Algebra may be found in the volume (Dejon and Henrici 1969) cited earlier. See in particular the article "A never failing fast convergent root-finding algorithm," by Bruno Dejon and Karl Nickel, pp. 1–35.

References

Argand, J.R.: Réflexions sur la nouvelle théorie d'analyse. Annales Math. **5**, 197–209 (1814)

Baltus, C.: Gauss's second proof of the fundamental theorem of algebra, 1815. Proc. Canad. Soc. Hist. Phil. Math. **13**, 97–106 (2000)

Baltus, C.: D'Alembert's proof of the fundamental theorem of algebra. Hist. Math. **31**, 414–428 (2004)

Courant, R., Robbins, H.: What is Mathematics? Oxford U., Oxford (1941)

D'Alembert, J.L.R.: Recherches sur le calcul intégral. Histoire de l'Academie Royale Berlin, 182–224 (1746)

Dejon, B., Henrici, P.: Constructive Aspects of the Fundamental Theorem of Algebra. Proceedings of a Symposium Conducted at the IBM Research Laboratory, Zürich-Rüschlkon, Switzerland, June 5–7, 1967. Wiley, New York (1969)

Dunham, W.: Euler and the fundamental theorem of algebra. College Math. J. **22**(4), 282–293 (1991)

Eaton, J.E.: The fundamental theorem of algebra. Amer. Math. Monthly **67**, 578–579 (1960)

Euler, L.: Recherches sur les racines imaginaires des equations. Mem. l'acad. sci. Berlin **5**, 222–288 (1749)

Fine, B., Rosenberger, G.: The Fundamental Theorem of Algebra. Springer, New York (1997)

Gauss, C.F.: Die vier Gauss'schen Beweise für die Zerlegung ganzer algebraischer Funktionen in reelle Factoren ersten oder zweiten Grades (1799–1849), trans. E. Netto. Ostwalds Klassiker der exakten Wissenschaften 14. Wilhelm Engelmann, Leipzig (1890)

---

[26] In his later paper (Smale 1981) Smale analyzed algorithms for finding roots of polynomials from the perspective of computational complexity theory, where questions of speed of convergence are paramount.

[27] The authors assert without proof that "in fact it is easy to see that the [full] sequence $\{z_j\}$ converges to a root" of p.

Gilain, C.: Sur l'histoire du theéorème d'algèbre: theorie des équations et calcul intégral. Arch. Hist. Exact Sci. **42**(2), 91–136 (1991)

Gordan, P.: Ueber den Fundamentalsatz der Algebra. Math. Ann. **10**, 572–575 (1876)

Hirsch, M., Smale, S.: On algorithms for solving $f(x) = 0$. Comm. Pure App. Math. **32**, 281–312 (1979)

Kline, M.: Mathematical Thought from Ancient to Modern Times. Oxford U., Oxford (1972)

Kneser, H.: Der Fundamentalsatz der Algebra und Intuitionismus. Math. Z. **46**, 287–302 (1940)

Kneser, M.: Ergänzung zu einer Arbeit von Hellmuth Kneser über den Fundamentalsatz der Algebra. Math. Z. **177**, 285–287 (1981)

Lax, P.D., Zalcman, L.: Complex Proofs of Real Theorems. University Lecture Series 58. Amer. Math. Soc., Providence (2012)

Ostrowski, A.: Über den ersten und vierten Gaussschen Beweise des Fundamentalsatzes der Algebra. In Alexander Ostrowski, Collected Mathematical Papers, vol. 1, pp. 538–553. Birkhäuser, Basel et al. (1983)

Remmert, R.: The fundamental theorem of algebra. In H-D. Ebbinghaus (ed.), Numbers, pp. 97–122. Springer, New York (1990)

Sen, A.: Fundamental theorem of algebra—yet another proof. Amer. Math. Monthly **107**(9), 842–843 (2000)

Smale, S.: The fundamental theorem of algebra and complexity theory. Bull. Amer. Math. Soc. **4**(1), 1–36 (1981)

Smithies, F.: A forgotten paper on the fundamental theorem of algebra. Notes Rec. Royal Soc. London **54**(3), 333–341 (2000)

Weyl, H.: Randbemerkungen zu Hauptproblemen der Mathematik. Math. Z. **20**, 131–150 (1924)

Wood, J.: On the roots of equations. Philos. Trans. Royal Soc. London **88**, 368–377 (1798)

Zalcman, L.: Picard's theorem without tears. Amer. Math. Monthly **85**, 265–268 (1978)

Chapter 9
Desargues's Theorem

The basic notions, and some of the fundamental theorems, of what would later be called projective geometry were first established in the seventeenth century, in response to questions that arose with regard to map projections and problems of perspective encountered by artists in representing three-dimensional scenes on planar canvases. Viewed then as theorems of *Euclidean* geometry, many of the results obtained could not be expressed in full generality by a single statement, but required consideration of various special cases.

An important example of such a result is **Desargues's Theorem in the Plane**, first stated and proved by Girard Desargues in a privately circulated manuscript in 1639.[1] In Euclidean terms it may be expressed by the following set of statements: Let $A_1 B_1 C_1$ and $A_2 B_2 C_2$ be two triangles in a plane, with $A_1 \neq A_2$, $B_1 \neq B_2$, and $C_1 \neq C_2$. If the lines through A_1 and A_2, B_1 and B_2 and C_1 and C_2 are all distinct and either meet at a point O or are all parallel, then:

(D1) If the extensions of each pair of corresponding sides of the two triangles intersect, the points of intersection are collinear.

(D2) If two pairs of corresponding sides of the triangles are parallel, the third pair of corresponding sides are also parallel.

(D3) If one pair of corresponding sides of the triangles are parallel and the extensions of the other two pairs of corresponding sides intersect, the points of intersection lie on a line parallel to the two parallel sides.

Figures 9.1–9.3 illustrate the various types of configurations that can arise. (See also Crannell and Douglas 2012.) The three diagrams in Figure 9.1 all represent configurations of type (D1) in which the lines through corresponding vertices intersect at O. They differ in the position of the point O relative to the two triangles: between them in (a), outside both in the same direction in (b), and inside one

---

[1]For more on the history of Desargues's Theorem, see section 14.3 in Kline (1972) and the illuminating case study in Arana and Mancosu (2012).

J.W. Dawson, Jr., *Why Prove it Again?*, DOI 10.1007/978-3-319-17368-9_9

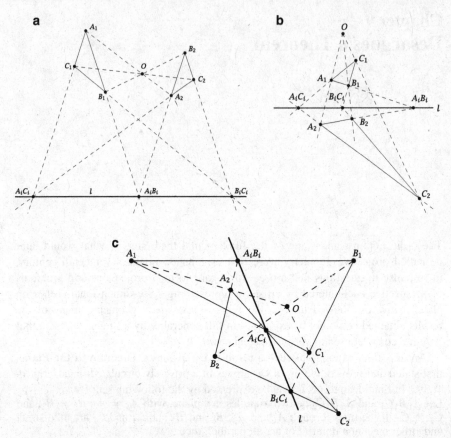

Fig. 9.1 (a) Case (D1), O between both triangles. (b) Case (D1), O on the same side of both triangles. (c) Case (D1), O inside one triangle

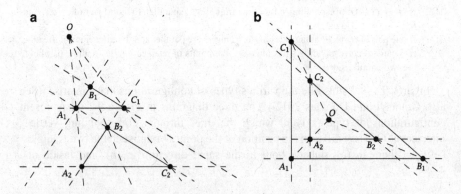

Fig. 9.2 (a) Case (D2), O outside both triangles. (b) Case (D2), O inside both triangles

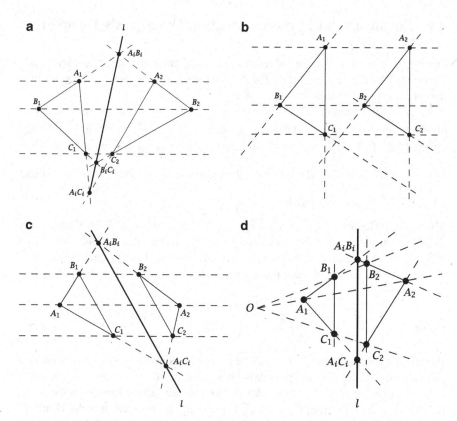

Fig. 9.3 (a) Case (D1), parallel projection. (b) Case (D2), parallel projection. (c) Case (D3), parallel projection. (d) Case (D3), projection from external point

triangle but outside the other in (c). Figure 9.2 likewise shows configurations of type (D2) in which the lines through corresponding vertices intersect at O, located outside both triangles in 9.2(a) and inside both in 9.2(b). Configurations in which the lines through corresponding vertices are all parallel are shown in Figures 9.3(a)–(c). Diagram 9.3(a) illustrates type (D1), diagram 9.3(b) type (D2), and diagram 9.3(c) type (D3). Finally, diagram 9.3(d) illustrates type (D3) when the lines through corresponding points intersect. (Note: In the configurations of types (D1) and (D3), $A_i B_i$ designates the point of intersection of lines $\overleftrightarrow{A_1 B_1}$ and $\overleftrightarrow{A_2 B_2}$, and similarly for $A_i C_i$ and $B_i C_i$.)

9.1   Euclidean and Cartesian proofs of Desargues's Theorem

Desargues derived (D1) from the result of classical antiquity known as Menelaus's
Theorem, which gives a criterion for three points that lie either on the sides of a
triangle or on their extensions to be collinear.

Menelaus's Theorem is most conveniently stated in terms of the lengths of
directed line segments. Accordingly, given points A and B on an (arbitrarily)
directed line L, let $|\overrightarrow{AB}|$ denote the directed distance along L from A to B. Then
$|\overrightarrow{BA}| = -|\overrightarrow{AB}|$, and if A, B, and C are three distinct points on L, $\dfrac{|\overrightarrow{AC}|}{|\overrightarrow{CB}|} > 0$ if and
only if C lies between A and B.

Menelaus's Theorem: Given a triangle ABC, let $D, E,$ and F be three points
distinct from the vertices, which lie, respectively, on the sides $\overline{BC}, \overline{AC},$ and $\overline{AB}$ or
their extensions, on each of which a positive direction has been arbitrarily assigned.
Then $D, E,$ and F are collinear if and only if

$$\frac{|\overrightarrow{AF}|}{|\overrightarrow{FB}|} \cdot \frac{|\overrightarrow{BD}|}{|\overrightarrow{DC}|} \cdot \frac{|\overrightarrow{CE}|}{|\overrightarrow{EA}|} = -1.$$

Proof: Suppose first that D, E, F all lie on line L. Since a line that intersects one
side of a triangle but does not pass through a vertex must intersect exactly one other
side, either two of D, E, F lie on sides (rather than their extensions), or none do. In
the first case, by relabeling the vertices if necessary, we may assume that D and F
are the points on sides. In either case, draw the line through A parallel to L, and let
P be its point of intersection with $\overline{BC}$ (Figure 9.4(a)) or its extension (Figure 9.4(b)).
Then (1) triangle ABP is similar to triangle FBD and (2) triangle ACP is similar to
triangle ECD, so

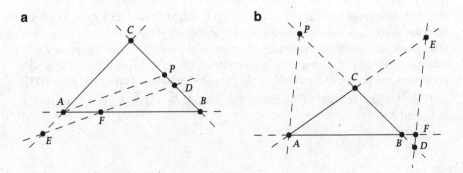

a   **b**

Fig. 9.4 Proof of Menelaus's Theorem

$$\frac{|\overrightarrow{AF}|}{|\overrightarrow{FB}|} = \frac{|\overrightarrow{DP}|}{|\overrightarrow{BD}|} = -\frac{|\overrightarrow{PD}|}{|\overrightarrow{BD}|} \quad \text{(by (1))} \quad \text{and} \quad \frac{|\overrightarrow{CE}|}{|\overrightarrow{EA}|} = -\frac{|\overrightarrow{CD}|}{|\overrightarrow{PD}|} = \frac{|\overrightarrow{DC}|}{|\overrightarrow{PD}|} \quad \text{(by (2))}.$$

Therefore

$$\frac{|\overrightarrow{AF}|}{|\overrightarrow{FB}|} \cdot \frac{|\overrightarrow{BD}|}{|\overrightarrow{DC}|} \cdot \frac{|\overrightarrow{CE}|}{|\overrightarrow{EA}|} = -\frac{|\overrightarrow{PD}|}{|\overrightarrow{BD}|} \cdot \frac{|\overrightarrow{BD}|}{|\overrightarrow{DC}|} \cdot \frac{|\overrightarrow{DC}|}{|\overrightarrow{PD}|} = -1.$$

Conversely, suppose by *reductio* that

$$\frac{|\overrightarrow{AF}|}{|\overrightarrow{FB}|} \cdot \frac{|\overrightarrow{BD}|}{|\overrightarrow{DC}|} \cdot \frac{|\overrightarrow{CE}|}{|\overrightarrow{EA}|} = -1,$$

but that D, E, F are not collinear. Then either all three factors in that product are negative, or just one is; that is, either none of D, E, F are on the sides of triangle ABC or exactly two of them are. In the latter case, we may assume as before that D and F are the two on the sides. Since E is not between A and C, the line L through D and E is not parallel to $\overline{AB}$. Let G be its point of intersection with $\overline{AB}$. Then D, E, and G are collinear, so by the 'only if' part of the theorem proved above,

$$\frac{|\overrightarrow{AG}|}{|\overrightarrow{GB}|} \cdot \frac{|\overrightarrow{BD}|}{|\overrightarrow{DC}|} \cdot \frac{|\overrightarrow{CE}|}{|\overrightarrow{EA}|} = -1.$$

In particular, $\dfrac{|\overrightarrow{AG}|}{|\overrightarrow{GB}|} > 0$, so G must lie between A and B (Figure 9.5(a)). Comparison with the *reductio* assumption then shows that

$$\frac{|\overrightarrow{AG}|}{|\overrightarrow{GB}|} = \frac{|\overrightarrow{AF}|}{|\overrightarrow{FB}|},$$

whence

$$\frac{|\overrightarrow{AB}|}{|\overrightarrow{GB}|} = \frac{|\overrightarrow{AG}|}{|\overrightarrow{GB}|} + \frac{|\overrightarrow{GB}|}{|\overrightarrow{GB}|} = \frac{|\overrightarrow{AF}|}{|\overrightarrow{FB}|} + \frac{|\overrightarrow{FB}|}{|\overrightarrow{FB}|} = \frac{|\overrightarrow{AB}|}{|\overrightarrow{FB}|}.$$

Therefore $|\overrightarrow{GB}| = |\overrightarrow{FB}|$, so $G = F$. Thus D, E, F must be collinear, contrary to assumption.

If none of D, E, F lie on the sides of triangle ABC, the same proof works unless $\overline{DE}$ is parallel to $\overline{AB}$ (Figure 9.5(b)). In that case, the line through E and F must intersect the extension of $\overline{AB}$, say at G, and we may argue similarly that

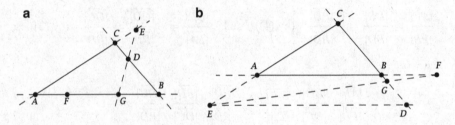

Fig. 9.5 Proof of the converse to Menelaus's Theorem

$$\frac{|\overrightarrow{AF}|}{|\overrightarrow{FB}|} \cdot \frac{|\overrightarrow{BG}|}{|\overrightarrow{GC}|} \cdot \frac{|\overrightarrow{CE}|}{|\overrightarrow{EA}|} = -1.$$

By the *reductio* assumption,

$$\frac{|\overrightarrow{BD}|}{|\overrightarrow{DC}|} = -\frac{|\overrightarrow{DB}|}{|\overrightarrow{DC}|} = -\frac{|\overrightarrow{GB}|}{|\overrightarrow{GC}|} = \frac{|\overrightarrow{BG}|}{|\overrightarrow{GC}|}, \text{ so}$$

$$\frac{|\overrightarrow{BC}|}{|\overrightarrow{DC}|} = \frac{|\overrightarrow{DC}|}{|\overrightarrow{DC}|} - \frac{|\overrightarrow{DB}|}{|\overrightarrow{DC}|} = \frac{|\overrightarrow{GC}|}{|\overrightarrow{GC}|} - \frac{|\overrightarrow{GB}|}{|\overrightarrow{GC}|} = \frac{|\overrightarrow{BC}|}{|\overrightarrow{GC}|}.$$

Then $|\overrightarrow{DC}| = |\overrightarrow{GC}|$, so $D = G$, once again contradicting the assumption that D, E, F are not collinear.   q.e.d.

Using Menelaus's Theorem, statement (D1) may now be proved as follows: Consider any of the diagrams of Figure 9.1, and for notational convenience, let $P = B_i C_i, Q = A_i C_i$, and $R = A_i B_i$. Then triangle $B_1 C_1 O$ and the line through B_2, C_2 and P satisfy the hypotheses of Menelaus's Theorem, as do triangle $A_1 C_1 O$ and the line through A_2, C_2, and Q, and triangle $A_1 B_1 O$ and the line through A_2, B_2, and R. Hence

$$\frac{|\overrightarrow{B_1 P}|}{|\overrightarrow{PC_1}|} \cdot \frac{|\overrightarrow{C_1 C_2}|}{|\overrightarrow{C_2 O}|} \cdot \frac{|\overrightarrow{OB_2}|}{|\overrightarrow{B_2 B_1}|} = -1,$$

$$\frac{|\overrightarrow{C_1 Q}|}{|\overrightarrow{QA_1}|} \cdot \frac{|\overrightarrow{A_1 A_2}|}{|\overrightarrow{A_2 O}|} \cdot \frac{|\overrightarrow{OC_2}|}{|\overrightarrow{C_2 C_1}|} = -1, \quad \text{and}$$

$$\frac{|\overrightarrow{A_1 R}|}{|\overrightarrow{RB_1}|} \cdot \frac{|\overrightarrow{B_1 B_2}|}{|\overrightarrow{B_2 O}|} \cdot \frac{|\overrightarrow{OA_2}|}{|\overrightarrow{A_2 A_1}|} = -1.$$

The product of these three equations (after canceling like terms, recalling that $|\overrightarrow{BA}| = -|\overrightarrow{AB}|$) reduces to

$$\frac{|\overrightarrow{B_1 P}|}{|\overrightarrow{PC_1}|} \cdot \frac{|\overrightarrow{C_1 Q}|}{|\overrightarrow{QA_1}|} \cdot \frac{|\overrightarrow{A_1 R}|}{|\overrightarrow{RB_1}|} = -1.$$

By the converse to Menelaus's Theorem, P, Q and R are thus collinear.

To prove (D2), let sides $\overline{A_1 B_1}$ and $\overline{A_2 B_2}$ be parallel and sides $\overline{A_1 C_1}$ and $\overline{A_2 C_2}$ be parallel. Suppose first that the lines through A_1 and A_2, B_1 and B_2, and C_1 and C_2 are concurrent at O (Figure 9.2(a)). Then

$$\frac{|\overline{OB_1}|}{|\overline{B_1 B_2}|} = \frac{|\overline{OA_1}|}{|\overline{A_1 A_2}|} = \frac{|\overline{OC_1}|}{|\overline{C_1 C_2}|},$$

where the first equality holds because $\overline{A_1 B_1}$ and $\overline{A_2 B_2}$ are parallel and the second because $\overline{A_1 C_1}$ and $\overline{A_2 C_2}$ are parallel. The equality of the first and third terms shows that the segments $\overline{B_1 C_1}$ and $\overline{B_2 C_2}$ divide the segments $\overline{OB_2}$ and $\overline{OC_2}$ proportionally. Hence $\overline{B_1 C_1}$ and $\overline{B_2 C_2}$ are parallel.

If, on the other hand, the lines through A_1 and A_2, B_1 and B_2, and C_1 and C_2 are all parallel (Figure 9.3(b)), then $A_1 A_2 B_1 B_2$ and $A_1 A_2 C_1 C_2$ are parallelograms, so $|\overline{A_1 B_1}| = |\overline{A_2 B_2}|$ and $|\overline{A_1 C_1}| = |\overline{A_2 C_2}|$. Also, the angles $B_1 A_1 C_1$ and $B_2 A_2 C_2$ must be equal. Consequently the triangles $A_1 B_1 C_1$ and $A_2 B_2 C_2$ are congruent by the side-angle-side criterion. Therefore sides $\overline{B_1 C_1}$ and $\overline{B_2 C_2}$ make equal angles with the parallel sides $\overline{A_1 B_1}$ and $\overline{A_2 B_2}$, so they must also be parallel.

Statement (D3) for the case of parallel projection is readily proven by Cartesian analytic means. Without loss of generality, we may presume that the projection lines are horizontal, as in Figure 9.3(c), and that the y-axis passes through A_1; say the equations for $\overleftrightarrow{A_1 A_2}$, $\overleftrightarrow{B_1 B_2}$ and $\overleftrightarrow{C_1 C_2}$ are $y = 0$, $y = a$ and $y = b$, respectively. Let A_2 have coordinates $(x_3, 0)$, B_1 have coordinates (x_1, a), and C_1 have coordinates (x_2, b). Then since $\overline{B_1 C_1}$ and $\overline{B_2 C_2}$ are parallel by hypothesis, $|\overline{B_1 B_2}| = |\overline{C_1 C_2}| = d$, so B_2 has coordinates $(x_1 + d, a)$ and C_2 has coordinates $(x_2 + d, b)$. The equations for lines $\overleftrightarrow{A_1 B_1}$ and $\overleftrightarrow{A_2 B_2}$ are then $y = \dfrac{a}{x_1}x$ and $y = \dfrac{a}{x_1 - x_3 + d}$, so their intersection $A_i B_i$ has coordinates $\left(\dfrac{x_1 x_3}{x_3 - d}, \dfrac{a x_3}{x_3 - d} \right)$. Similarly, the equations for $\overleftrightarrow{A_1 C_1}$ and $\overleftrightarrow{A_2 C_2}$ are $y = \dfrac{b}{x_2}x$ and $y = \dfrac{b}{x_2 - x_3 + d}$, so they intersect at $\left(\dfrac{x_2 x_3}{x_3 - d}, \dfrac{b x_3}{x_3 - d} \right)$.

If $x_1 \neq x_2$, the slope of line l is then $\dfrac{a - b}{x_1 - x_2}$, which is the same as that of lines $\overleftrightarrow{B_1 C_1}$ and $\overleftrightarrow{B_2 C_2}$. If $x_1 = x_2$, then $\overleftrightarrow{B_1 C_1}$ and $\overleftrightarrow{B_2 C_2}$ are vertical lines whose equations are $x = x_1$ and $x = x_1 + d$, and the x-coordinates of both $A_i B_i$ and $A_i C_i$ satisfy $x = \dfrac{x_1 x_3}{x_3 - d}$ and so must be identical. Therefore line l is also vertical.

It remains to prove (D3) for the case of projection from a point in the plane. So, as in Figure 9.3(d), suppose that triangles $A_1 B_1 C_1$ and $A_2 B_2 C_2$ are perspective from

O, that sides $\overline{B_1C_1}$ and $\overline{B_2C_2}$ are parallel, and that lines $\overleftrightarrow{A_1B_1}$ and $\overleftrightarrow{A_2B_2}$ intersect at A_iB_i and lines $\overleftrightarrow{A_1C_1}$ and $\overleftrightarrow{A_2C_2}$ at A_iC_i. It is to be shown that the line through A_iB_i and A_iC_i is parallel to $\overline{B_1C_1}$. To simplify the notation, relabel A_iB_i as D. Invoking the Parallel Postulate, let l be the line through D that *is* parallel to $\overline{B_1C_1}$, and let l intersect $\overleftrightarrow{A_1C_1}$ at E. Then it suffices to show that E coincides with A_iC_i.

Toward that end, note that triangles DB_1B_2 and EC_1C_2 are in *parallel* perspective (cf. Figure 9.3(a)). Hence by (D1), proven above using Menelaus's Theorem, the extensions of corresponding sides of those two triangles must intersect in collinear points. Now $\overleftrightarrow{B_1B_2}$ and $\overleftrightarrow{C_1C_2}$ intersect at O and $\overleftrightarrow{DB_1}$ and $\overleftrightarrow{EC_1}$ intersect at A_1, so $\overleftrightarrow{DB_2}$ and $\overleftrightarrow{EC_2}$ must intersect at a point on $\overleftrightarrow{OA_1}$. But $\overleftrightarrow{DB_2}$ intersects $\overleftrightarrow{OA_1}$ at A_2. Thus A_2 is on line $\overleftrightarrow{EC_2}$. That is, $\overleftrightarrow{A_2C_2}$ passes through E, which lies on $\overleftrightarrow{A_1C_1}$. Therefore $\overleftrightarrow{A_1C_1}$ and $\overleftrightarrow{A_2C_2}$ must intersect at E.   q.e.d.

9.2 Desargues's Theorem in a projective context

Given such a plethora of different configurations satisfying the hypotheses of (D1)—(D3), a concise, uniform statement of Desargues's Theorem might seem unattainable. But as first recognized by Johannes Kepler in 1604, and thirty-five years later by Desargues himself, unification can be achieved by adjoining to each set of parallel lines in the Euclidean plane an ideal *point at infinity*, the totality of such points constituting the *line at infinity*. The points of that extended structure constitute the (real) *projective plane*.

Any two distinct parallel lines in the Euclidean plane are regarded as concurrent in the projective plane at the corresponding point at infinity. Consequently, in the projective plane Desargues's Theorem may be stated much more succinctly. It is customary in that regard to say that two triangles in the Euclidean plane whose three pairs of corresponding vertices lie on distinct concurrent lines in the projective plane are *perspective from a point*, and that two triangles in the Euclidean plane whose three pairs of corresponding sides intersect in collinear points of the projective plane are *perspective from a line*. In those terms the statement of the theorem then becomes

Desargues's Theorem in the Projective Plane: If two triangles in the Euclidean plane are perspective from a point in the projective plane, they are perspective from a line in the projective plane.

Recasting Desargues's Theorem in projective terms thus enables clauses (D1)–(D3) of the Euclidean version to be unified into a single statement. More importantly, however, it makes possible a uniform *proof* of the various cases described in those clauses.

Homogeneous coordinates in the projective plane

More formally, the projective plane may be defined as the set of all equivalence classes of ordered triples (x, y, z) of real numbers, excluding $(0, 0, 0)$, with respect to the equivalence relation given by $(x, y, z) \equiv (mx, my, mz)$ for all real numbers $m \neq 0$. The equivalence classes may be regarded as providing homogeneous *coordinates* for the points of the projective plane.

Such coordinates may be interpreted in more than one way. If the Euclidean plane P is embedded in $\mathbb{R}^3$ one Cartesian unit above the origin O (as, e.g., in Courant and Robbins (1941), pp. 193–194), each point A of that plane may be identified with the line OA through O and A and be given the homogeneous coordinates $(ma, mb, mc), m \neq 0$, corresponding to the Cartesian direction vectors $\langle ma, mb, mc \rangle$ of that line, while each ideal point I may be given coordinates $(ma, mb, 0), m \neq 0$ and be identified with the line through O parallel to P with Cartesian direction vectors $\langle ma, mb, 0 \rangle$. The points of the projective plane are thus placed in one-to-one correspondence with the lines through the origin in $\mathbb{R}^3$.

For the purpose of proving Desargues's Theorem, however, it is more convenient (as in Coxeter 1962, pp. 234–237) to adopt Möbius's idea of interpreting (x, y, z) as representing the point with *barycentric* coordinates $(\mu_A x, \mu_B y, \mu_C z)$ relative to a fixed Euclidean reference triangle $A1B_1C_1$ bearing non-zero weights μ_A, μ_B, μ_C at its vertices. By definition, $A_1 = (1, 0, 0)$, $B_1 = (0, 1, 0)$, and $C_1 = (0, 0, 1)$. The weights μ_A, μ_B, μ_C serve as parameters; by adjusting their values, any point none of whose coordinates are zero with respect to the reference triangle can be made to have coordinates $(1, 1, 1)$.

In terms of such barycentric coordinates, every line is represented by a *homogeneous* linear equation. The barycentric equation for the line through two given points (a, b, c) and (d, e, f) is

$$\begin{vmatrix} a & b & c \\ d & e & f \\ x & y & z \end{vmatrix} = 0.$$

In particular, the lines extending the sides $\overline{B_1C_1}, \overline{A_1C_1}$, and $\overline{A_1B_1}$ of the reference triangle have the barycentric equations $x = 0$, $y = 0$, and $z = 0$.

The following argument[2] then provides a direct, uniform proof of Desargues's Theorem.

Analytic Proof of Desargues's Theorem in the Projective Plane:

Suppose triangles $A_1B_1C_1$ and $A_2B_2C_2$ in the Euclidean plane are perspective from a point O in the projective plane, as in any of figures 9.1–9.3. Take $A_1B_1C_1$ to be the reference triangle. Then none of the coordinates of O can be zero. (For if the first coordinate were zero, O, B_1 and C_1 would all lie on the same line, and hence so would B_2 and C_2. Therefore the lines extending sides $\overline{B_1C_1}$ and $\overline{B_2C_2}$ would

---

[2] A more detailed version of that given at http://planetmath.org/proofofdesarguestheorem.

coincide, contrary to hypothesis. Likewise, if the second coordinate of O were zero, the extensions of sides $\overline{A_1C_1}$ and $\overline{A_2C_2}$ would coincide, and if the third coordinate of O were zero, the extensions of sides $\overline{A_1B_1}$ and $\overline{A_2B_2}$ would coincide.) So, as noted above, the weights at A_1, B_1, and C_1 may be chosen so that O has coordinates $(1, 1, 1)$.

The equations for the lines through O and A_1, O and B_1, and O and C_1 are then given by

$$\begin{vmatrix} 1 & 1 & 1 \\ 1 & 0 & 0 \\ x & y & z \end{vmatrix} = 0, \quad \begin{vmatrix} 1 & 1 & 1 \\ 0 & 1 & 0 \\ x & y & z \end{vmatrix} = 0, \quad \text{and} \quad \begin{vmatrix} 1 & 1 & 1 \\ 0 & 0 & 1 \\ x & y & z \end{vmatrix} = 0,$$

which reduce to $y = z$, $x = z$ and $x = y$, respectively. As points on those lines, A_2, B_2, and C_2 must then have coordinates of the forms (x, z, z), (z, y, z), and (y, y, z); and since the coordinates are homogeneous, there must be constants a, b, c such that $A_2 = (1, a, a)$, $B_2 = (b, 1, b)$, and $C_2 = (c, c, 1)$.

The equations for the lines through A_2 and B_2, B_2 and C_2, A_2 and C_2 are thus

$$\begin{vmatrix} 1 & a & a \\ b & 1 & b \\ x & y & z \end{vmatrix} = 0, \quad \begin{vmatrix} b & 1 & b \\ c & c & 1 \\ x & y & z \end{vmatrix} = 0, \quad \text{and} \quad \begin{vmatrix} 1 & a & a \\ c & c & 1 \\ x & y & z \end{vmatrix} = 0;$$

that is,

$$(ab - a)x + (ab - b)y + (1 - ab)z = 0,$$

$$(1 - bc)x + (bc - b)y + (bc - c)z = 0, \quad \text{and}$$

$$(a - ac)x + (ac - 1)y + (c - ac)z = 0.$$

The intersection point A_iB_i must satisfy the equations $z = 0$ and $(ab - a)x + (ab - b)y = 0$; B_iC_i must satisfy $x = 0$ and $(bc - b)y + (bc - c)z = 0$; and A_iC_i must satisfy $y = 0$ and $(a - ac)x + (c - ac)z = 0$. Since the coordinates are homogeneous, the value of one of the variables in the second of each pair of equations may be specified and the other be determined from the equation. Taking $y = ab - a$ in the first pair gives $x = b - ab$; taking $z = bc - b$ in the second pair gives $y = c - bc$; and taking $z = a - ac$ in the third gives $x = ac - c$. Coordinates for A_iB_i, B_iC_i and A_iC_i are then $(b - ab, ab - a, 0)$, $(0, c - bc, bc - b)$ and $(ac - c, 0, a - ac)$. They will be collinear if and only if

$$\begin{vmatrix} b - ab & ab - a & 0 \\ 0 & c - bc & bc - b \\ ac - c & 0 & a - ac \end{vmatrix} = 0.$$

A straightforward computation confirms that that is so.

9.3   Desargues's Theorem in three-dimensional space

In addition to proving the planar statement (D1) and its converse, Desargues also proved the following result.

Desargues's Theorem in $\mathbb{R}^3$**:** If two triangles lying in different, non-parallel Euclidean planes in $\mathbb{R}^3$ are perspective from a point O in $\mathbb{R}^3$, they are perspective from a line in $\mathbb{R}^3$.

Surprisingly, a Euclidean proof of the spatial correlate of statement (D1), in the case of projection from a point not at infinity, is almost immediate. (See Figure 9.6.)

For if the line $\overleftrightarrow{A_1 B_1}$ in plane P_1 intersects the line $\overleftrightarrow{A_2 B_2}$ in plane P_2 at the point $A_i B_i$, $A_i B_i$ must lie on the line of intersection of those two planes, and similarly for $A_i C_i$ and $B_i C_i$.

The same argument also establishes the spatial correlate of (D3), since if, say, line $\overleftrightarrow{A_1 B_1}$ in plane P_1 is parallel to line $\overleftrightarrow{A_2 B_2}$ in P_2, those two lines must each be parallel to the line of intersection of P_1 and P_2. The spatial correlate of (D2), on the other hand, is vacuously true, since if triangles $A_1 B_1 C_1$ in P_1 and $A_2 B_2 C_2$ in P_2 have two pairs of corresponding sides parallel, the planes containing those two triangles must be parallel.

Desargues's Theorem in $\mathbb{R}^3$ can then be employed (as in Gans (1969), pp. 258–259) to give a **spatial geometric proof** of the corresponding planar statement (D1). (See Figure 9.7.)

Thus, suppose triangles $A_1 B_1 C_1$ and $A_2 B_2 C_2$ lie in a plane E in $\mathbb{R}^3$ and are perspective from a point O in E. Suppose also that no pair of corresponding sides of those triangles are parallel. It is to be shown that the three points of intersection of those pairs of corresponding sides are collinear.

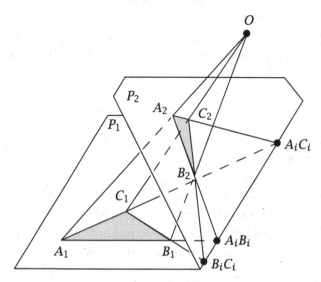

Fig. 9.6 Desargues's Theorem in $\mathbb{R}^3$ (adapted from Courant and Robbins (1941))

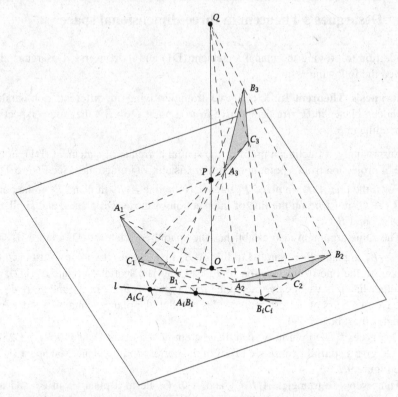

Fig. 9.7 Desargues's Theorem in the Plane as a corollary of Desargues's Theorem in $\mathbb{R}^3$

Toward that end, let P be a point in $\mathbb{R}^3$ not in E, and let Q be any point beyond P on the ray from O through P. Consider the line $\overleftrightarrow{A_1P}$ through A_1 and P. Since O lies on the line $\overleftrightarrow{A_1A_2}$ through A_1 and A_2, the ray through O and P, and hence the point Q, lie in the plane determined by the lines $\overleftrightarrow{A_1P}$ and $\overleftrightarrow{A_1A_2}$. Consequently the line through A_2 and Q also lies in that plane, and must intersect the non-parallel line $\overleftrightarrow{A_1P}$ at some point A_3. Similarly, lines $\overleftrightarrow{B_1P}$ and $\overleftrightarrow{B_2Q}$ must intersect at some point B_3, and lines $\overleftrightarrow{C_1P}$ and $\overleftrightarrow{C_2Q}$ must intersect at some point $C3$.

The points A_3, B_3 and C_3 so determined project through P onto the non-collinear points A_1, B_1 and C_1, and so are themselves non-collinear. That is, the triangles $A_1B_1C_1$ and $A_3B_3C_3$ are perspective from point P. Likewise, the triangles $A_2B_2C_2$ and $A_3B_3C_3$ are perspective from point Q.

Since the line $\overleftrightarrow{A_3C_3}$ lies in the plane of triangle A_1C_1P and is not parallel to $\overleftrightarrow{A_1C_1}$, it must intersect $\overleftrightarrow{A_1C_1}$ at a point on the line of intersection of E and the plane of triangle A_1C_1P. Similarly, $\overleftrightarrow{A_3C_3}$ lies in the plane of triangle A_2C_2Q and is not parallel to $\overleftrightarrow{A_2C_2}$, so it must intersect the latter at a point on the line of intersection of E and the plane of triangle $A_2 C_2 Q$. But the line $\overleftrightarrow{A_3 C_3}$ cannot intersect the plane E in more than one point. So the extensions of sides $\overline{A_1 C_1}, \overline{A_2 C_2},$ and $\overline{A_3 C_3}$ must all intersect at a common point $A_i C_i$.

In the same way, the extensions of sides $\overline{B_1 C_1}, \overline{B_2 C_2}$ and $\overline{B_3 C_3}$ must all intersect at $B_i C_i$, and the extensions of sides $\overline{A_1 B_1}, \overline{A_2 B_2},$ and $\overline{A_3 B_3}$ must all intersect at $A_i B_i$. That is, the corresponding sides of the coplanar triangles $A_1 B_1 C_1$ and $A_2 B_2 C_2$ intersect at the same three points as the corresponding sides of the triangles $A_1 B_1 C_1$ and $A_3 B_3 C_3$, which lie in non-parallel planes. By Desargues's Theorem in $\mathbb{R}^3$, the latter must be collinear.

Another spatial proof of Desargues's Theorem in the Plane (outlined, e.g., in Courant and Robbins (1941), pp. 187–188) proceeds from the same assumptions — that two triangles in a plane E that is embedded in $\mathbb{R}^3$ are perspective from a point O of E not at infinity, and that P is a point of $\mathbb{R}^3$ not in E. But the conclusion (that the triangles are perspective from a line in E) is deduced not from Desargues's Theorem in $\mathbb{R}^3$, but by reducing statement (D1) to statement (D2), which was proved above by an elementary synthetic argument.

To do so, the plane E containing the given Desarguesian configuration is projected from P onto a plane H that is *parallel* to the plane determined by P and two of the three points of intersection (say $A_i B_i$ and $B_i C_i$) of corresponding sides of the given triangles. (See Figure 9.8, an orthographic projection of E onto H, based on the configuration in Figure 9.1(a), in which the point P is taken to lie on the line through $A_i B_i$ that is parallel to H and perpendicular to line l.) In such a projection, the line l through $A_i B_i$ and $B_i C_i$ is projected onto the line at infinity, and for any point q on l, all the lines in E that pass through q are projected onto lines in H that are *parallel* to the line through q and P.

In particular, A_1 and B_1 will be projected onto points A_1' and B_1' on a line in H that is parallel to the line through P and $A_i B_i$, and A_2 and B_2 will be projected onto points A_2' and B_2' on another such line. Likewise, B_1 and C_1 will be projected onto points on a line in H parallel to the line through P and $B_i C_i$, and B_2 and C_2 onto another such line. And since projection preserves not only lines but incidence relations, the triples of points $A_1' B_1' C_1'$ and $A_2' B_2' C_2'$ will be perspective from the point O' in H that is the image of O under the projection from P.

Then *if* the segment $\overline{A_1 B_1}$ projects onto the segment $\overline{A_1' B_1'}$, the segment $\overline{A_2 B_2}$ onto $\overline{A_2' B_2'}$, and similarly for all the other segments that form the sides of triangles $A_1 B_1 C_1$ and $A_2 B_2 C_2$, the triangles $A_1' B_1' C_1'$ and $A_2' B_2' C_2'$ will have two corresponding sides parallel. So by (D2), proven earlier,[3] the third pair of corresponding sides, $\overleftrightarrow{A_1' C_1'}$ and $\overleftrightarrow{A_2' C_2'}$, must also be parallel, which can only happen if the point of intersection $A_i C_i$ in E of $\overleftrightarrow{A_1 C_1}$ and $\overleftrightarrow{A_2 C_2}$ lies on line l. q.e.d

---

[3] That proof referred to Figure 9.2(a), in which the point O lay on the same side of triangles $A_1 B_1 C_1$ and $A_2 B_2 C_2$. If O lies between those triangles, as in Figure 9.8, it is easy to check that the same reasoning still applies.

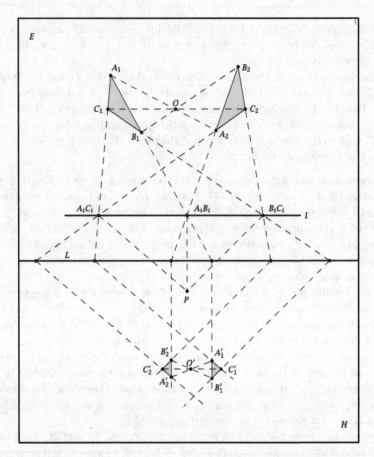

Fig. 9.8 Orthographic projection of Figure 9.1(a) onto a plane parallel to line l

Such will be the case so long as line l does not intersect any of the sides of triangles $A_1B_1C_1$ or $A_2B_2C_2$. If it *does*, as in the configuration shown in Figure 9.1(c), any side intersected by l, say $\overline{A_1B_1}$, will project not onto the segment $\overline{A_1'B_1'}$ along some line in H, but onto the *complement* of that segment along that line. The image of triangle $A_1B_1C_1$ will then not be a triangle at all, but will consist of two disjoint figures, each infinite in extent.[4]

The argument just given thus establishes statement (D1) whenever triangles $A_1B_1C_1$ and $A_2B_2C_2$ both lie on the same side of the line l (as, for example, in Figure 9.1(a)), or when l passes between them (as in Figure 9.1(b)). In the former

---

[4]That difficulty is passed over in silence in Courant and Robbins (1941) and other sources I have seen that sketch this argument. It is discussed in detail in Crannell and Douglas (2012), which illustrates how difficult such projections can be to visualize.

case, their projections will be triangles that lie on the same side of the line of intersection of planes E and H; in the latter case, one of those triangles will lie on each side of that line.

9.4 Comparative summary

The four proofs of Desargues's Theorem in the Plane presented above differ in their generality, reflect different strategies, and draw upon different methodologies: The first proof employs synthetic methods (Menelaus's Theorem and Euclidean similarity theory) to establish statements (D1) and (D2), and Cartesian analytic computations to deduce (D3) for the case of parallel projection. The proof in terms of projective coordinates establishes all of statements (D1)–(D3) analytically, without having to distinguish cases. The third proof establishes statement (D1), for the case of projection from a point, by projecting a spatial Desarguesian configuration onto a planar one. And the last proof establishes that same result (for some but not all configurations) by transforming a point-projective Desarguesian configuration of type (D1) into a corresponding parallel-projective configuration of type (D2).

9.5 Axiomatics and questions of purity

Because the statement of Desargues's Theorem in the Plane involves only the projective notions of incidence, concurrence and collinearity, questions may be (and historically, were) raised concerning the purity of the methods used to prove it. As a *geometric* statement, one may prefer a synthetic proof to an analytic one. As a *planar* theorem, one may object to employing spatial considerations in its proof. Or, as a *projective* assertion, one may desire to prove it without introducing metric notions such as lengths of segments or similarity of triangles. To what extent can these various demands be reconciled?

That question only makes sense within the context of a specific axiomatic framework. For in formalizing projective geometry one might, after all, take the statement of Desargues's Theorem in the Plane, or of some other purely projective planar statement that implies it, to be an *axiom*.[5] In Coxeter (1962), for example, Desargues's Theorem is derived by an argument first given in Hessenberg (1905),

---

[5]Such an axiomatization serves a two-fold purpose: it allows projective geometry to be developed on a purely projective basis, and it demonstrates the primacy of particular theorems in that development. As such, it exemplifies still another reason for giving an alternative proof: **to demonstrate dependencies among different mathematical statements** and so determine what principles are essential for the derivation of a given result (the central aim of the so-called 'reverse mathematics' program in mathematical logic).

as a consequence of five axioms for the general projective plane, the last of which is what is known in Euclidean geometry as Pappus's Theorem: that if the six vertices of a hexagon lie alternately on two lines, the three points of intersection of pairs of opposite sides are collinear.[6] But of course, neither Desargues's Theorem nor Pappus's Theorem is self-evident. So the question becomes: Can Desargues's Theorem in the Plane be proved in a more 'natural' axiomatization of projective geometry?

One such axiomatization was given by Guiseppe Peano. In Peano (1894) he gave fifteen postulates, all but the last of which were planar, and asserted that "[Desargues's Theorem] in the plane is . . . a consequence of postulate XV, and thus . . . of solid geometry." But from a theorem of Eugenio Beltrami he concluded that Desargues's Theorem in the Plane was *not* a consequence of his other postulates. For Beltrami had proved that the only smooth surfaces capable of planar representations in which each point of the surface corresponded to one of the plane and each geodesic to a straight line were those of constant curvature; and Peano claimed that although Desargues's Theorem would hold on any surface of constant curvature that satisfied the first fourteen of his postulates, it would not necessarily hold on a surface of *non*constant curvature that satisfied them. He did not, however, give an explicit example of such a surface.[7]

In his *Grundlagen der Geometrie* (Hilbert 1899), David Hilbert gave another axiomatization for geometry, intended to remedy the gaps in Euclid's presentation; and in that system he showed definitively that any proof of Desargues's Theorem in the Plane must either invoke spatial axioms or metric notions. There were five groups of axioms in Hilbert's system: incidence axioms, the first two of which were planar; order axioms; the parallel axiom; congruence axioms; and continuity axioms. Hilbert noted that Desargues's Theorem in the Plane could be proved from either the spatial incidence axioms or from (all) the congruence axioms, but he gave a model (illustrated in Figure 9.9, based on Figure 13 in Arana and Mancosu 2012) that satisfied the planar incidence axioms, the parallel axiom, all of the order and continuity axioms, and the first five of the congruence axioms, in which Desargues's Theorem in the Plane did not hold for some Desarguesian configurations.

Specifically, Hilbert considered an ellipse centered at the origin of the Euclidean plane, with horizontal major axis of length one and vertical minor axis of length one half. He imagined the ellipse as acting like a lens, refracting line segments that crossed it into arcs of circles. In particular, the segment joining points p and q on the ellipse would become the part of the circle passing through p, q and $(3/2, 0)$ that lay within the ellipse. Lines in the model were then taken to be all Euclidean straight

---

[6]In Euclidean geometry, Pappus's Theorem, like Desargues's, is a consequence of Menelaus's Theorem. It implies but is not implied by Desargues's Theorem.

[7]One example, an ellipsoid of revolution, has recently been given by Patrick Popescu-Pampu. It is described in detail on pp. 318–320 of Arana and Mancosu (2012), an article that treats many of the issues discussed in this section in much greater depth.

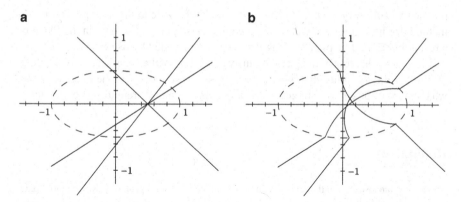

Fig. 9.9 (a) Incident lines before "refraction". (b) The same lines after "refraction"

lines that did not intersect the ellipse, together with all the curves that resulted from straight lines that did intersect the ellipse when the segment of them within the ellipse was replaced by the corresponding 'refracted' arc.

Figure 9.9(a) shows three Euclidean lines that, before 'refraction,' are incident at a point within the ellipse, while Figure 9.9(b) illustrates how the refracted lines no longer intersect at a common point. It is this possibility that causes Desargues's Theorem to fail in the model.

Hilbert went further, however: for in Theorem 35 of the *Grundlagen* he also showed that in any geometry that satisfies the planar incidence axioms, the order axioms, the parallel axiom *and* Desargues's Theorem in the Plane, it is possible to construct an algebra of segments that forms an ordered skew field, from which in turn a model of *all* the incidence axioms together with the order axioms and the parallel axiom can be defined. Consequently, Desargues's Theorem in the Plane is both necessary and sufficient to ensure that a geometry that satisfies the planar incidence axioms, the order axioms and the parallel axiom can be embedded in a spatial geometry.

In that sense, Desargues's Theorem in the Plane has been considered by some to have "tacit spatial content" (see, e.g., Hallett 2008). For those who hold that view, spatial proofs of Desargues's Theorem in the Plane do not *ipso facto* violate purity of method. One may nevertheless object that, to the extent that it rests upon *algebraic* notions, Hilbert's proof of Theorem 35 is still impure. However, an alternative proof of Hilbert's embedding theorem that uses only geometric notions was given by F.W. Levy in 1939, and another, quite recently, by John Baldwin and William Howard. (See Levy 1939 and Baldwin 2013.)

Hilbert gave his own assessment of the significance of his embedding theorem in the notes for his lectures from 1898/99 (published in Hallett and Majer 2004): Having shown that a geometric proof of the planar Desargues's Theorem must invoke spatial considerations, he believed he had "for the first time ... put into practice *a critique of means of proof*" that gave substance to the notion of purity of method; for he had formally demonstrated that Desargues's Theorem in the Plane is a result that cannot be proved using just means "suggested by its content."

In the next chapter the question of purity of method will arise again, in the context of number theory rather than geometry. In either case, however, the question as to what means of proof are "suggested by a theorem's content" remains a vexing one.[8]

References

Arana, A., Mancosu, P.: On the relationship between plane and solid geometry. Rev. Symb. Logic **5**(2), 249–353 (2012)

Baldwin, J.: Formalization, primitive concepts, and purity. Rev. Symb. Logic **6**(1), 87–128 (2013)

Courant, R., Robbins, H.: What is Mathematics? Oxford U.P., Oxford (1941)

Coxeter, H.S.M.: Introduction to Geometry. Wiley, New York and London (1962)

Crannell, A., Douglas, S.: Drawing on Desargues. Math. Intelligencer **34**(2), 7–14 (2012)

Gans, D.: Transformations and Geometries. Appleton-Century-Crofts, New York (1969)

Hallett, M.: Reflections on the purity of method in Hilbert's *Grundlagen der Geometrie*. In P. Mancosu (ed.), The Philosophy of Mathematical Practice, pp. 198–255. Oxford U.P., Oxford (2008)

Hallett, M., Majer, U. (eds.): David Hilbert's Lectures on the Foundations of Geometry, 1891–1902. Springer-Verlag, Berlin (2004)

Hessenberg, G.: Beweis des Desarguesschen Satzes aus dem Pascalschen. Math. Annalen **16**, 161–172 (1905)

Hilbert, D.: Grundlagen der Geometrie. B.G. Teubner, Leipzig (1899)

Kline, M.: Mathematical Thought from Ancient to Modern Times. Oxford U.P., Oxford (1972)

Levy, F.: On a fundamental theorem of geometry. J. Indian Math. Soc. **3–4**, 82–92 (1939)

Peano, G.: Sui fondamenti della geometria. Rivista di Matematica **4**, 51–90 (1894)

---

[8]The papers of Arana and Mancosu, Baldwin, and Hallett cited above, and other papers cited therein, provide contrasting analyses of many of the issues in dispute, and are recommended as introductions to the burgeoning literature on the subject of methodologically pure proofs. Of particular interest is the contention defended in Baldwin (2013) — consonant with the viewpoint taken here, outlined in chapter 1 — that "purely formal methods" are themselves "impotent" for the purpose of characterizing the notion of purity of method.

Chapter 10
The Prime Number Theorem

In the wake of Euclid's proof of the infinitude of the primes, the question of how the primes were distributed among the integers became central — a question that has intrigued and challenged mathematicians ever since. The sieve of Eratosthenes provided a simple but very inefficient means of identifying which integers were prime, but attempts to find explicit, closed formulas for the nth prime, or for the number $\pi(x)$ of primes less than or equal to a given number x, proved fruitless. Eventually extensive tables of integers and their least factors were compiled, detailed examination of which suggested that the apparently unpredictable occurrence of primes in the sequence of integers nonetheless exhibited some statistical regularity. In particular, in 1792 Euler asserted that for large values of x, $\pi(x)$ was approximately given by $\dfrac{x}{\ln x}$; six years later, Legendre suggested $\dfrac{x}{\ln x - 1}$ and (wrongly) $\dfrac{x}{\ln x - 1.0836}$ as better approximations; and in 1849, in a letter to his student Encke (translated in the appendix to Goldstein 1973), Gauss mentioned his apparently long-held belief that the logarithmic integral

$$\mathrm{li}(x) = \int_2^x \frac{1}{\ln t}\, dt$$

gave a still better approximation.[1] Using the notation $f(x) \sim g(x)$ to denote the equivalence relation defined by $\lim_{x \to \infty} \dfrac{f(x)}{g(x)} = 1$, those conjectures may be expressed in asymptotic form by the statements

(PNT) $\qquad \pi(x) \sim \dfrac{x}{\ln x}, \quad \pi(x) \sim \dfrac{x}{\ln x - 1}, \quad$ and $\quad \pi(x) \sim \mathrm{li}(x).$

---

[1] Some texts instead define $\mathrm{li}(x)$ as $\lim_{\epsilon \to 0} \left(\int_0^{1-\epsilon} 1/\ln t\, dt + \int_{1+\epsilon}^x 1/\ln t\, dt \right)$, which adds a constant (approximately 1.04) to $\mathrm{li}(x)$ as defined above, but does not affect asymptotic arguments.

© Springer International Publishing Switzerland 2015
J.W. Dawson, Jr., *Why Prove it Again?*, DOI 10.1007/978-3-319-17368-9_10

Clearly $\dfrac{x}{\ln x} \sim \dfrac{x}{\ln x - 1}$, and integration by parts can be used to show that $\dfrac{x}{\ln x} \sim \mathrm{li}(x)$ as well. For

$$\mathrm{li}(x) = \int_2^x \frac{1}{\ln t}\, dt = K + \int_e^x \frac{1}{\ln t}\, dt = K + \frac{x}{\ln x} - e + \int_e^x \frac{1}{(\ln t)^2}\, dt$$

$$= (K - e) + \frac{x}{\ln x} + \left[\int_e^{\sqrt{x}} \frac{1}{(\ln t)^2}\, dt + \int_{\sqrt{x}}^x \frac{1}{(\ln t)^2}\, dt \right],$$

where K is a constant. Then since $1/(\ln t)^2$ is less than or equal to 1 on the interval $[e, \sqrt{x}]$ and less than or equal to $1/(\ln \sqrt{x})^2 = 4/(\ln x)^2$ on the interval $[\sqrt{x}, x]$,

$$\mathrm{li}(x) \leq (K - e) + \frac{x}{\ln x} + \sqrt{x} + \frac{4x}{(\ln x)^2}.$$

But also, $\mathrm{li}(x) \geq (x - 2)/\ln x$, since $1/\ln t$ is decreasing on $[2, x]$. Hence

$$\frac{x - 2}{x} \leq \frac{\ln x}{x}\,\mathrm{li}(x) \leq \frac{(K - e)\ln x}{x} + 1 + \frac{\ln x}{\sqrt{x}} + \frac{4}{\ln x}.$$

It follows immediately by L'Hopital's rule that $\lim_{x \to \infty} \dfrac{\ln x}{x}\,\mathrm{li}(x) = 1$; that is, $\mathrm{li}(x) \sim \dfrac{x}{\ln x}$.

The three statements (PNT) are thus logically equivalent. *Ipso facto*, however, they do not capture the full strength of the conjectures made by Legendre and Gauss, since they do not indicate how accurately each of the three formulas approximates $\pi(x)$. For that, estimates of the absolute or relative errors involved in those approximations are also needed.

The truth of those conjectures, together with error estimates, was finally established in 1896, independently and nearly simultaneously, by Jacques Hadamard and Charles de la Vallée Poussin, using techniques of complex contour integration discussed further below. Their long and complicated proofs of what has ever since been called the Prime Number Theorem were distinct, and have been analyzed in detail in Narkiewicz (2000). Later, proofs were devised that avoided some of the delicate issues involved in contour integration over infinite paths by invoking Norbert Wiener's Tauberian theory for Fourier integrals; but the arguments remained complex. Eventually, more than 80 years after the original proofs, Donald Newman found a short proof involving integration only over finite contours (Newman 1980); and in the meantime, against the expectation of most of the mathematical community, Atle Selberg and Paul Erdős (partly independently and again nearly simultaneously) published so-called 'elementary' proofs, involving no recourse to complex-analytic methods (Erdős 1949; Selberg 1949).[2]

---

[2]In part for those proofs, Selberg was awarded a Fields Medal in 1950 and Erdős the 1951 Cole Prize in Number Theory.

Because of the length and complexity of all but Newman's proof, the discussion that follows, unlike that in the preceding chapters, does not give full details of the various proofs, but rather focuses on their essential ideas and the differences among them. Readers seeking further details may consult the recent monograph (Jameson 2003), which provides a self-contained presentation of a descendant of the original proofs, as well as Newman's proof and an elementary proof based on that given in Levinson (1969), all presented at a level accessible to students who have had only basic courses in real and complex analysis. An overview of work on the Prime Number Theorem in the century since its first proofs is given in Bateman and Diamond (1996), while Diamond (1982) and Goldfeld (2004) discuss the history of elementary proofs of the theorem. Narkiewicz (2000) provides a comprehensive history of the overall development of prime number theory up to the time of Hardy and Littlewood, including alternative proofs of many results.

10.1 Steps toward the goal: Prior results of Dirichlet, Chebyshev, and Riemann

As noted in Narkiewicz (2000), p. 49, "the first use of analytic methods in number theory was made by P.G. Dirichlet" during the years 1837–39. In particular, Dirichlet proved that if k and l are relatively prime integers with $k < l$, then the arithmetic progression $\{nk + l\}$, where n ranges over the positive integers, must contain infinitely many primes. In doing so Dirichlet considered the series now named after him $\left(\text{those of the form } \sum_{n=1}^{\infty} \dfrac{f(n)}{n^s}\right)$ and gave an argument that could be adapted to show that if f is a completely multiplicative complex-valued function (that is, f is not identically zero and $f(mn) = f(m)f(n)$ for all integers m and n) and $\sum_{n=1}^{\infty} \dfrac{f(n)}{n^s}$ converges absolutely for some real value s_0, then it converges absolutely for all complex s with $\operatorname{Re} s \geq s_0$ and

$$(28) \qquad \sum_{n=1}^{\infty} \frac{f(n)}{n^s} = \prod_{p \text{ prime}} \sum_{j=0}^{\infty} \frac{f(p^j)}{p^{js}} = \prod_{p \text{ prime}} \frac{1}{1 - f(p)p^{-s}}.^3$$

With s restricted to real values, the special case $f(n) = 1$ is Euler's product formula (cf. Chapter 7, footnote 6):

$$(29) \qquad \zeta(s) = \sum_{n=1}^{\infty} \frac{1}{n^s} = \prod_{p \text{ prime}} \frac{1}{1 - p^{-s}}$$

(from which it follows that $\zeta(s) \neq 0$ for $\operatorname{Re} s > 1$).

---

[3] In fact, for any absolutely convergent series $\sum_{n=1}^{\infty} f(n)$ of non-zero terms, if $f(n)$ is completely multiplicative and no $f(n) = -1$, then $\sum_{n=1}^{\infty} f(n) = \prod_{p \text{ prime}} (1 - f(p))^{-1}$.

Nine years later, in the first of two papers on $\pi(x)$ (Chebyshev 1848), Pafnuty Chebyshev proved that for all integers $k > 1$ and any constant $C > 0$, there are infinitely many integers m and n for which

$$(30) \qquad \pi(m) < \text{li}(m) + C\,\frac{m}{\ln^k m} \quad \text{and} \quad \pi(n) > \text{li}(n) - C\,\frac{n}{\ln^k n}.$$

(For further details, see Narkiewicz (2000), pp. 98–102.)   Consequently,

$$\pi(m)\frac{\ln m}{m} < \text{li}(m)\frac{\ln m}{m} + \frac{C}{\ln^{k-1} m} \quad \text{and} \quad \pi(n)\frac{\ln n}{n} > \text{li}(n)\frac{\ln n}{n} - \frac{C}{\ln^{k-1} n}$$

for arbitrarily large integers m and n. Therefore, since $\text{li}(x) \sim \dfrac{x}{\ln x}$, if the $\lim_{x\to\infty} \dfrac{\pi(x)\ln(x)}{x}$ exists, then it must equal 1 $\left(\text{that is, } \pi(x) \sim \dfrac{x}{\ln x}\right)$.

In proving (30), Chebyshev made use of the Gamma function, defined for $\text{Re}\,s > 0$ by $\Gamma(s) = \int_0^\infty e^{-t} t^{s-1}\, dt$. In his second paper (Chebyshev 1850), however, he introduced two number-theoretic functions related to $\pi(x)$, namely

$$\theta(x) = \sum_{\substack{p\,prime \\ p \le x}} \ln p \quad \text{and} \quad \psi(x) = \sum_{\substack{p\,prime \\ p^n \le x}} \ln p = \sum_{n=1}^{m} \theta(x^{1/n}),$$

where m is the largest integer for which $2^m \le x$, and obtained important bounds on their values using only elementary means. In particular, since $\ln x$ is an increasing function, $\theta(x) \le \pi(x)\ln x \le x\ln x$ and $\psi(x) \le \pi(x)\ln x$; and since $m \le \ln x/\ln 2$,

$$\psi(x) - \theta(x) = \sum_{n=2}^{m} \theta(x^{1/n}) \le \theta(x^{1/2}) + m\theta(x^{1/3})$$

$$\le \frac{1}{2}x^{1/2}\ln x + \frac{\ln x}{\ln 2}\frac{1}{3}x^{1/3}\ln x$$

$$\le \frac{1}{2}x^{1/2}\ln x + \frac{\ln x}{\ln 2}\frac{x^{1/2}\ln x}{x^{1/6}}$$

$$\le 4x^{1/2}\ln x,$$

because the maximum value of $\dfrac{\ln x}{\ln 2}\dfrac{1}{x^{1/6}}$ occurs for $x = e^6$ and is less than 3.2. Thus $\psi(x) - 4\sqrt{x}\ln x \le \theta(x) \le \psi(x)$.

Hence if $\lim_{x\to\infty}\dfrac{\psi(x)}{x}$ exists, then so does $\lim_{x\to\infty}\dfrac{\theta(x)}{x}$, and they are equal. Furthermore, if $\lim_{x\to\infty}\dfrac{\theta(x)}{x}$ exists, then so must $\lim_{x\to\infty}\dfrac{\pi(x)\ln(x)}{x}$, because for $0 < \epsilon < 1$,

$$\theta(x) \ge \sum_{x^{1-\epsilon} \le p \le x} \ln p \ge (\pi(x) - \pi(x^{1-\epsilon}))(1-\epsilon)\ln x \ge (\pi(x) - x^{1-\epsilon})(1-\epsilon)\ln x;$$

so

$$\frac{\theta(x)}{x} \leq \frac{\pi(x) \ln x}{x} \leq \frac{1}{1-\epsilon} \frac{\theta(x)}{x} + \frac{\ln x}{x^\epsilon}.$$

As already noted, $\lim_{x \to \infty} \dfrac{\pi(x) \ln(x)}{x}$ must then equal 1. So to prove the Prime Number Theorem it suffices to prove that either $\lim_{x \to \infty} \dfrac{\psi(x)}{x}$ or $\lim_{x \to \infty} \dfrac{\theta(x)}{x}$ exists.

The principal theorem of Chebyshev (1850) was that there exist constants C_1, C_2, and C_3 such that for all $x \geq 2$,

(31)   (i) $C_1 x \leq \psi(x) \leq C_2 x$ and   (ii) $C_3 x \leq \theta(x) \leq C_2 x$.

Indeed, Chebyshev's methods (involving Stirling's formula) showed that C_2 can be taken to be 1.1224 and, if $x \geq 37$, that C_3 can be taken to be .73 (see Narkiewicz 2000, p. 111); and from the latter estimate Chebyshev deduced Bertrand's postulate (that for every $x > 1$ the interval $(x, 2x]$ contains some prime) as an immediate corollary. For direct calculation shows that Bertrand's postulate holds for $1 \leq x < 37$, and if it failed for some $x \geq 37$, the estimate would imply the contradiction that $1.46x \leq \theta(2x) = \theta(x) \leq 1.13x$.

Without using Stirling's formula, considerations involving binomial coefficients yield the weaker inequality $\theta(x) \leq x \ln 4 \approx 1.3863x$. The argument, given in Jameson (2003), p. 36, uses no number-theoretic facts beyond Euclid's lemma and its corollary that if distinct primes divide n, so does their product. Namely, for any given n the expansion of $(1+1)^{2n+1}$ includes the two equal binomial coefficients $\binom{2n+1}{n}$ and $\binom{2n+1}{n+1}$, so $\binom{2n+1}{n} < 2^{2n} = 4^n$. If $p_i, \ldots, p_{i+j}$ are all the primes between $n+2$ and $2n+1$, inclusive, then none of those primes divides the denominator $n!$ of $\binom{2n+1}{n}$, but each of them, and hence the product of all of them, divides its numerator, $n! \binom{2n+1}{n}$. Therefore that product must divide $\binom{2n+1}{n}$ itself and so must be less than $\binom{2n+1}{n}$. Consequently,

$$\theta(2n+1) - \theta(n+1) = \sum_{k=0}^{j} \ln p_{i+k} = \ln(p_i \cdots p_{i+j}) \leq \ln \binom{2n+1}{n} \leq n \ln 4.$$

Then since $\theta(n) \leq n \ln 4$ for $n = 2, 3, 4$, we may assume by induction on m that $\theta(k) \leq k \ln 4$ for $k \leq 2m$. The assumption holds for $m = 2$, and $m+1 < 2m$ for $m \geq 2$. So $\theta(m+1) \leq (m+1) \ln 4$, whence by the last displayed inequality, $\theta(2m+1) \leq (2m+1) \ln 4$; and since $2m+2$ is not prime, $\theta(2m+2) = \theta(2m+1)$. Thus $\theta(2m+2) \leq (2m+1) \ln 4 < (2m+2) \ln 4$, finishing the induction.

The inequality (31)(ii) also entails corresponding bounds on $\pi(x)$. For $\pi(x)$ can be expressed as the sum $\sum_{2 \leq n \leq x} (f(n)/\ln n)$, where $f(n) = \ln n$ if n is prime and 0 otherwise, and Abel summation[4] then gives

$$\pi(x) = \frac{\theta(x)}{\ln x} + \int_2^x \frac{\theta(t)}{t(\ln t)^2},$$

so (31)(ii)) yields

$$\frac{C_3 x}{\ln x} + C_3 \int_2^x \frac{1}{(\ln t)^2}\, dt \leq \pi(x) \leq \frac{C_2 x}{\ln x} + C_2 \int_2^x \frac{1}{(\ln t)^2}\, dt.$$

On the other hand, integration by parts shows that

$$\mathrm{li}(x) + \frac{2}{\ln 2} \doteq \frac{x}{\ln x} + \int_2^x \frac{1}{(\ln t)^2}\, dt.$$

Hence

$$C_3 \left(\mathrm{li}(x) + \frac{2}{\ln 2} \right) \leq \pi(x) \leq C_2 \left(\mathrm{li}(x) + \frac{2}{\ln 2} \right).$$

Similar arguments (see Jameson (2003), p. 35) show that for any $\epsilon > 0$, an x_0 must exist such that $(C_3 - \epsilon)\mathrm{li}(x) \leq \pi(x) \leq (C_2 - \epsilon)\mathrm{li}(x)$ for all $x > x_0$. The Prime Number Theorem would follow if it could be shown that the bounds in Chebyshev's estimate can be taken to be $C_2 = C_3 = 1$. But Chebyshev's methods were inadequate for that task.

A key breakthrough came a decade later in Bernhard Riemann's memoir "Ueber die Anzahl der Primzahlen unter einer gegebener Größe" (Riemann 1860). Riemann began by recalling Euler's product formula (29), but took s therein to be a *complex* number and initially defined $\zeta(s)$ to be the function of s given by the expressions on each side of Euler's formula whenever both expressions converged — that is, when $\mathrm{Re}\, s > 1$. So defined, $\zeta(s)$ is analytic on the half plane $\mathrm{Re}\, s > 1$, but Riemann went on immediately to extend it to a function analytic on all of $\mathbb{C}$ except $s = 1$. To do so he employed the Gamma function (actually, the function $\Pi(s) = \Gamma(s+1)$) together with the substitution $t = nx$ to obtain

$$\frac{\Gamma(s)}{n^s} = \int_0^\infty e^{-nx} x^{s-1}\, dx$$

---

[4]The result of applying the formula

$$\sum_{2 \leq n \leq x} a(n)g(n) = \left[\sum_{n \leq x} a(n) \right] g(x) - \int_2^x \left[\sum_{n \leq t} a(n) \right] g'(t)\, dt,$$

valid whenever $a(1) = 0$ and $g(x)$ has a continuous derivative on $[2, x]$.

as the nth term of the geometric series for $\Gamma(s)\zeta(s)$. Then

$$\Gamma(s)\zeta(s) = \sum_{n=1}^{\infty} \int_0^{\infty} e^{-nx} x^{s-1}\, dx = \int_0^{\infty} x^{s-1} \sum_{n=1}^{\infty} e^{-nx}\, dx = \int_0^{\infty} \frac{x^{s-1}}{e^x - 1}\, dx.$$

For $s \neq 1$ he equated $e^{-\pi s i} - e^{\pi s i}$ times the latter integral to the contour integral

$$\int_C \frac{(-x)^{s-1}}{e^x - 1}\, dx,$$

"taken from $+\infty$ to $+\infty$ in a positive sense" around a region containing "in its interior the point 0 but no other point of discontinuity of the integrand." He concluded that $\zeta(s)$ could then be defined for all $s \neq 1$ by the equation

(32) $$2 \sin(\pi s)\Gamma(s)\zeta(s) = i \int_C \frac{(-x)^{s-1}}{e^x - 1}\, dx,$$

and that so defined $\zeta(s)$ would be analytic except for a simple pole at $s = 1$; that is, $\zeta(s) - 1/(s-1)$ would be an entire function.

Riemann noted that for $\mathrm{Re}\, s < 0$ the same integral could be taken in the reverse direction, surrounding the region exterior to the curve C. It would then be "infinitely small for all s of infinitely large modulus," and the integrand would be discontinuous only at the points $2n\pi i$. Its value would thus be $\sum_{n=1}^{\infty}(-n2\pi i)^{s-1}(-2\pi i)$, whence

$$2 \sin(\pi s)\Gamma(s)\zeta(s) = (2\pi)^s \sum_{n=1}^{\infty} n^{s-1}((-i)^{s-1} + i^{s-1}) = (2\pi)^s \zeta(1-s)2 \sin\frac{\pi s}{2},$$

which reduces to the functional equation

$$\zeta(1-s) = \frac{2}{(2\pi)^s} \cos(\frac{\pi s}{2})\Gamma(s)\zeta(s).$$

In the remainder of his memoir, Riemann made some assertions concerning $\pi(x)$ and the non-real zeros of the zeta function — assertions equivalent to the following statements:

(i) The number $N(T)$ of non-real roots ρ of $\zeta(s)$ for which $|\mathrm{Im}\, \rho| \leq T$ (with multiple roots counted according to their multiplicity) is asymptotically equal to

(33) $$\frac{T}{2\pi} \ln \frac{T}{2\pi} - \frac{T}{2\pi}.$$

(ii) It is "very likely" that all non-real zeros of $\zeta(s)$ satisfy $\mathrm{Re}\, s = 1/2$.

(iii) If $F(x) = 1/2k + \sum_{n=1}^{\infty} \dfrac{\pi(x^{1/n})}{n}$ whenever $x = p^k$ for some prime p and

$F(x) = \sum_{n=1}^{\infty} \dfrac{\pi(x^{1/n})}{n}$ otherwise, then for $x > 1$,

$$F(x) = \mathrm{li}(x) - \sum_{\rho} \mathrm{li}(x^{\rho}) + \int_{x}^{\infty} \frac{dx}{(x^2 - 1)x \ln x} - \ln 2,$$

where ρ ranges over all non-real roots of $\zeta(s)$.

Assertions (i) and (iii) were later proved rigorously in von Mangoldt (1895), while the conjecture that all non-real zeros of $\zeta(s)$ satisfy $\operatorname{Re} s = 1/2$ is the still unproven Riemann Hypothesis.

Beyond the specific results it contained, three aspects of Riemann's memoir were seminal for the subsequent proofs of the Prime Number Theorem: the idea of regarding the series $\sum_{n=1}^{\infty} n^{-s}$ as a function of a *complex* argument; the extension of that function to a function analytic in the region $s \neq 1$, to which Cauchy's theory of contour integration could be applied; and the revelation that the distribution of prime numbers was intimately related to the location of the non-real zeros of that extended zeta function.[5]

10.2 Hadamard's proof

Just a year after receiving his degree of Docteur ès Sciences, Jacques Hadamard published the first of a series of papers concerning properties of the Riemann zeta function. The series culminated in his paper Hadamard (1896b), in which he proved the Prime Number Theorem in the form $\lim_{x \to \infty} \dfrac{\theta(x)}{x} = 1$.

Hadamard began that paper by noting that $\zeta(s)$ was analytic except for a simple pole at $s = 1$, and though it had infinitely many zeros with real part between 0 and 1 (a consequence of results in his earlier paper Hadamard 1893), it was non-zero for all s with $\operatorname{Re} s > 1$. His first goal was then to show that $\zeta(s)$ was also non-zero for all s with $\operatorname{Re} s = 1$.

To establish that he considered $\ln(\zeta(s))$, which by Euler's product theorem and the Maclaurin series for $\ln(1 - x)$ is equal to $\sum_{k=1}^{\infty} \sum_{p \ prime} \dfrac{1}{kp^{ks}}$. He split that sum into two parts, so that

$$(34) \quad \ln(\zeta(s)) = \sum_{p \ prime} \frac{1}{p^s} + \sum_{k=2}^{\infty} \sum_{p \ prime} \frac{1}{kp^{ks}} = S(s) + \sum_{k=2}^{\infty} \sum_{p \ prime} \frac{1}{kp^{ks}},$$

---

[5]A detailed commentary on developments stemming from Riemann's classic paper is given in Edwards (1974).

and noted that since $(s - 1)\zeta(s)$ is an entire function, it is bounded in any closed neighborhood of $s = 1$. Thus in any such neighborhood the difference $\ln(\zeta(s)) - \ln \dfrac{1}{s-1} = \ln((s - 1)\zeta(s))$ is a bounded function, so if s approaches 1 from the right along the real axis, $\ln(\zeta(s))$ must approach $+\infty$. But by (34), in any sufficiently small neighborhood of $s = 1$, $\ln(\zeta(s))$ also differs from $S(s)$ by a bounded expression, since the second term on the right of (34) is analytic for $\operatorname{Re} s > 1/2$. So $S(s)$ also approaches $+\infty$ as s approaches 1 from the right along the real axis.

On the other hand, consider $\ln(|\zeta(s)|)$. By (34), for $s = \sigma + it$, $\ln(|\zeta(s)|) = \operatorname{Re} \ln(\zeta(s))$ differs from

$$P(s) = \sum_{p\ prime} \operatorname{Re} p^{-s} = \sum_{p\ prime} \frac{\cos(t \ln p)}{p^\sigma}$$

by a bounded function; and if $\zeta(1 + it_0) = 0$ for some $t_0 \neq 0$, then $1 + it_0$ must be a simple root of $\zeta(s)$, whence $\ln(|\zeta(\sigma + it)|)$ must differ from $\ln(\sigma - 1)$ by a bounded function. So $\ln(|\zeta(\sigma + it_0)|)$ and $P(\sigma + it_0)$ must both approach $-\infty$ as σ approaches 1 from the right, and $P(\sigma + it_0)$ must differ from $-S(\sigma + it_0)$ by a bounded function.

Now suppose $0 < \rho = 1 - \epsilon < 1$, $0 < \alpha < 1$, and let P_n and S_n denote the partial sums of P and S for $p \leq n$. Hadamard divided the set of primes $p \leq n$ into two subsets, according to whether or not, for some integer k, the prime p satisfied the inequality $|t_0 \ln p - (2k + 1)\pi| < \alpha$. Writing $S_n = S_n' + S_n''$ and $P_n = P_n' + P_n''$ to correspond to that division, he used elementary inequalities to conclude that there must exist an integer N_ρ such that for all $n \geq N_\rho$, $\rho_n(\sigma) = S_n'(\sigma + it_0)/S_n(\sigma + it_0) > \rho$, since otherwise $P(\sigma + it_0) \geq -\theta S(\sigma)$, where $\theta = 1 - \epsilon + \epsilon \cos \alpha < 1$. Then, by the result of the previous paragraph, for some function F *bounded* throughout the half-plane $\operatorname{Re} s > 1/2$, $-S(\sigma + it_0) + F(\sigma + it_0) \geq -\theta S(\sigma)$. That is, $F(\sigma + it_0) \geq (1 - \theta)S(\sigma)$, with $1 - \theta > 0$. But as noted above, $\lim_{\sigma \to 1+} S(\sigma) = +\infty$.

The argument just given, which rests on the assumption that $\zeta(1 + it_0) = 0$, applies to any ρ satisfying $0 < \rho < 1$. If ρ is further required to satisfy $\dfrac{1}{1 + \cos(2\alpha)} < \rho < 1$, then similar manipulations of inequalities show that for $n \geq N_\rho$ and $s = \sigma + i(2t_0)$,

$$P_n(s) \geq \rho_n S_n(\sigma) \cos(2\alpha) + (\rho_n - 1)S_n(\sigma)$$
$$> \rho S_n(\sigma) \cos(2\alpha) + (\rho - 1)S_n(\sigma) = \Theta S_n(\sigma),$$

with $\Theta = \rho[(1 + \cos(2\alpha)) - 1] > 0$. Hence $P(s) = \lim_{n \to +\infty} P_n(s) \geq \Theta S(\sigma)$. Once again, $\lim_{\sigma \to 1} S(\sigma) = +\infty$, so $P(\sigma + i(2t_0))$, and thus also $\ln |\zeta(\sigma + i(2t_0))|$, must approach $+\infty$ as σ approaches 1 from the right. But that cannot be, since $\zeta(s)$ is analytic at the point $1 + i(2t_0)$. The assumption that $\zeta(1 + it_0) = 0$ is thereby refuted.

For the rest of his proof Hadamard drew upon ideas of E. Cahen, who in his doctoral dissertation at the École Normale Supérieure had unsuccessfully attempted to prove the Prime Number Theorem.

Given real numbers a and x, with $0 < x \neq 1$, Cahen had considered the contour integrals

$$\frac{1}{2\pi i} \int_{a-\infty i}^{a+\infty i} \frac{x^z}{z} \, dz, \quad \text{for } a > 0, \quad \text{and} \quad -\frac{1}{2\pi i} \int_{a-\infty i}^{a+\infty i} \frac{x^z}{z} \frac{\zeta'(z)}{\zeta(z)} \, dz, \quad \text{for } a > 1.$$

Hadamard considered instead the integrals

$$\frac{1}{2\pi i} \int_{a-\infty i}^{a+\infty i} \frac{x^z}{z^\mu} \, dz, \quad \text{for } a > 0, \quad \text{and} \quad -\frac{1}{2\pi i} \int_{a-\infty i}^{a+\infty i} \frac{x^z}{z^\mu} \frac{\zeta'(z)}{\zeta(z)} \, dz, \quad \text{for } a > 1,$$

where $\mu > 0$, which he denoted by J_μ and ψ_μ, respectively.

In order to evaluate the integrals J_μ, Hadamard distinguished three cases: $\mu =$ an integer, n; μ is non-integral and $x < 1$; or μ is non-integral and $x > 1$. For his proof of the Prime Number Theorem, however, only the first case was needed (indeed, just the case $n = 2$), which he established by integrating by parts $n - 1$ times, using the identity $1/z^n = \dfrac{(-1)^{n-1}}{(n-1)!} \dfrac{d^{n-1}}{dz^{n-1}}(1/z)$.[6] That gave

$$J_n = \frac{1}{2\pi i} \int_{a-\infty i}^{a+\infty i} \frac{x^z}{z^n} \, dz = \frac{1}{2\pi i} \frac{(\ln x)^{n-1}}{(n-1)!} \int_{a-\infty i}^{a+\infty i} \frac{x^z}{z} \, dz,$$

whence

$$(35) \qquad J_n = \begin{cases} 0 & \text{if } x < 1, \\ \dfrac{(\ln x)^{n-1}}{(n-1)!} & \text{if } x > 1, \end{cases}$$

since von Mangoldt had shown the year before that

$$(36) \qquad \frac{1}{2\pi i} \int_{a-\infty i}^{a+\infty i} \frac{x^z}{z} \, dz = \begin{cases} 0 & \text{if } x < 1, \\ 1/2 & \text{if } x = 1, \quad \text{(von Mangoldt 1895).} \\ 1 & \text{if } x > 1, \end{cases}$$

---

[6]In the other cases Hadamard used the identity $1/z^\mu = \dfrac{(-1)^{\mu-1}}{\Gamma(\mu)} \dfrac{d^{\mu-1}}{dz^{\mu-1}}(1/z)$, together with Cauchy's integral theorem, to obtain the general formula

$$J_\mu = \frac{1}{2\pi i} \int_{a-\infty i}^{a+\infty i} \frac{x^z}{z^\mu} \, dz = \begin{cases} 0 & \text{if } x < 1, \\ \dfrac{(\ln x)^{\mu-1}}{\Gamma(\mu)} & \text{if } x > 1. \end{cases}$$

To evaluate the integrals ψ_μ, Hadamard noted that by (29), $\ln(\zeta(s)) = -\sum_p \ln(1 - p^{-s})$, so logarithmic differentiation gives

$$(37) \qquad \frac{\zeta'(s)}{\zeta(s)} = -\sum_p \frac{(\ln p) p^{-s}}{1 - p^{-s}} = -\sum_p \ln p \sum_{k=1}^{\infty} \frac{1}{p^{ks}} = -\sum_{k=1}^{\infty} \frac{\Lambda(k)}{k^s},$$

where Λ denotes the von Mangoldt function, defined by $\Lambda(k) = \ln p$ if $k = p^m$ for some prime p and integer m and $\Lambda(k) = 0$ otherwise.

Then for $\mu = 2$,

$$\psi_2(x) = -\frac{1}{2\pi i} \int_{a-\infty i}^{a+\infty i} \frac{\zeta'(s)}{\zeta(s)} \frac{x^s}{s^2} \, ds$$

$$= \sum_p \ln p \frac{1}{2\pi i} \int_{a-\infty i}^{a+\infty i} \sum_{k=1}^{\infty} \frac{x^s}{p^{ks}} \frac{1}{s^2}$$

$$= \sum_p \ln p \sum_{k=1}^{\infty} \frac{1}{2\pi i} \int_{a-\infty i}^{a+\infty i} \frac{(x/p^k)^s}{s^2},$$

where by (35) the integrals in the last member are equal to 0 if $x/p^k < 1$ (that is, if $x^{1/k} < p$) and equal to $\ln\left(\dfrac{x}{p^k}\right)$ otherwise. Hence

$$\psi_2(x) = \sum_{k=1}^{\infty} \sum_{p \leq x^{1/k}} \ln p \ln\left(\frac{x}{p^k}\right) = \sum_{p \leq x} \ln p \ln\left(\frac{x}{p}\right) + \sum_{k=2}^{\infty} \sum_{p \leq x^{1/k}} \ln p \ln\left(\frac{x}{p^k}\right).$$

The double sum in the last term of the equation above is only apparently infinite, since the inner sum is vacuous for $k > \ln x / \ln 2$. Thus finally

$$(38) \qquad \psi_2(x) = \sum_{p \leq x} \ln p \ln\left(\frac{x}{p}\right) + \sum_{k=2}^{[\frac{\ln x}{\ln 2}]} \sum_{p \leq x^{1/k}} \ln p \ln\left(\frac{x}{p^k}\right),$$

where the brackets above the penultimate summation symbol denote the greatest integer function.

In the second term of (38) the first summation involves no more than $\ln x / \ln 2$ summands and the second summation no more than $\sqrt{x}$, the largest of which is that for $k = 2$. Consequently,

$$\psi_2(x) \leq \sum_{p \leq x} \ln p \ln\left(\frac{x}{p}\right) + \sqrt{x} \frac{\ln x}{\ln 2} \ln(\sqrt{x}) \ln x = \sum_{p \leq x} \ln p \ln\left(\frac{x}{p}\right) + \sqrt{x} \frac{\ln^3 x}{2 \ln 2}.$$

When divided by x, the last term above approaches 0 as x approaches ∞. Hadamard's final goals were then to show

(i) that $\lim_{x\to\infty} \frac{1}{x} \sum_{p\leq x} \ln p \ln\left(\frac{x}{p}\right) = 1$

and

(ii) that (i) implies that $\lim_{x\to\infty} \frac{\theta(x)}{x} = 1$.

Hadamard established (ii) via elementary but rather involved $\epsilon - \delta$ computations, showing that for every $\epsilon > 0$, $x - 2\epsilon x \leq \theta(x) \leq x + \frac{5\epsilon}{2}x$. (For details, see Narkiewicz 2000, pp. 202–204.) To prove (i) he showed that $\lim_{x\to\infty} \frac{\psi_2(x)}{x} = 1$, using Cauchy's integral theorem, von Mangoldt's theorem justifying Riemann's asymptotic estimate (33) of the number $N(T)$ of non-real roots ρ of $\zeta(s)$ for which $|\mathrm{Im}\,\rho| \leq T$, and two results from his earlier paper Hadamard (1896a)— in particular,

(A) If $\rho_1, \rho_2, \ldots$ are the non-real roots of $\zeta(s)$ ordered according to increasing absolute value, then $\sum_n 1/|\rho_n|^2$ converges.

and

(B) If $s \neq 1$ is not a root of $\zeta(s)$, then for some constant K

$$(39) \qquad \frac{\zeta'(s)}{\zeta(s)} = \frac{1}{1-s} + \sum_\rho \left(\frac{1}{s-\rho} + \frac{1}{\rho}\right) + K,$$

where ρ ranges over all roots of $\zeta(s)$, ordered according to increasing absolute value.

It follows from (A) that for any $\epsilon > 0$ there is an integer M such that $\sum_{n>M} 1/|\rho_n|^2 < \epsilon$. Let I be the maximum of $|\mathrm{Im}\,\rho_n|$ for $n \leq M$ and R the maximum of $|\mathrm{Re}\,\rho_n|$ for $n \leq M$. Since $\zeta(s)$ has no roots ρ with $\mathrm{Re}\,\rho \geq 1$, $R < 1$. Then, given $a > 1$, to compute

$$\psi_2(x) = -\frac{1}{2\pi i} \int_{a-\infty i}^{a+\infty i} \frac{\zeta'(s)}{\zeta(s)} \frac{x^s}{s^2}\, ds$$

Hadamard considered the infinite family of polygons $\Gamma = \mathbf{ABGECDFHA}$ defined in terms of a real parameter y as follows (see Figure 10.1): Choose $d > I$ such that no root of $\zeta(s)$ lies on the line $\mathrm{Im}(s) = d$, take c to be a real number satisfying $R < c < 1$, and let $\mathbf{C}$ be the point $c + di$ and $\mathbf{D}$ the point $c - di$. For $y > d$, let $\mathbf{A}$ be the point $a - yi$ and $\mathbf{B}$ the point $a + yi$. Finally, fix $e < 0$, let $\mathbf{E}$ be the point $e + di$ and $\mathbf{F}$ the point $e - di$, and denote by $\mathbf{G}$ and $\mathbf{H}$ the points where the lines $\mathrm{Im}(s) = \pm y$ intersect the lines from the origin that pass through $\mathbf{E}$ and $\mathbf{F}$.

Fig. 10.1 Contour used
by Hadamard to evaluate
the integral

$$\frac{1}{2\pi i}\int_{a-\infty i}^{a+\infty i}\frac{\zeta'(z)}{\zeta(z)}\frac{x^z}{z^2}\,dz$$

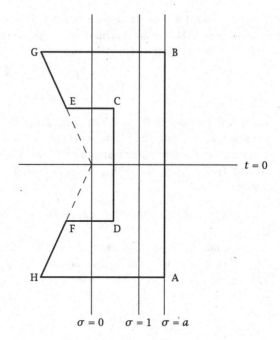

Hadamard used (33) and (39) to show that the integral of $\dfrac{\zeta'(s)}{\zeta(s)}\dfrac{x^s}{s^2}$ along each of the segments **BG** and **AH** approaches 0 as y approaches ∞.

In particular, he concluded from (33) that for any $A > 1$ and any positive integer λ the number of roots ρ of $\zeta(s)$ for which $|\text{Im}\,\rho|$ lies between $A^{3\lambda+3}$ and $A^{3\lambda}$ does not exceed $K\lambda A^{3\lambda}$, where K is a constant. The same bound applies *a fortiori* to the number of roots N_r for which $|\text{Im}\,\rho|$ lies between $3\lambda + 1$ and $3\lambda + 2$. On the other hand, the asymptotic expression for $N(T)$ given in (33) approaches ∞ as λ does, so for sufficiently large λ there must be two consecutive roots ρ_j and ρ_{j+1} whose imaginary parts β_j and β_{j+1} differ by at least $\dfrac{A^{3\lambda+2} - A^{3\lambda+1}}{K\lambda A^{3\lambda}} = \dfrac{A(A-1)}{K\lambda}$. Consequently, for $y_\lambda = (\beta_j + \beta_{j+1})/2$, any root of $\zeta(s)$ must lie above or below the line $\text{Im}(s) = y_\lambda$ by at least $\dfrac{A(A-1)}{2K\lambda}$.

That being so, if **BG** lies along $\text{Im}(s) = y_\lambda$, the summation in (39) may be estimated by splitting it into two parts, the first sum ranging over all roots ρ satisfying $A^{3\lambda} \le \text{Im}\,\rho \le A^{3\lambda+3}$ and the second over all other roots. The definition of y_λ entails that the first sum is bounded by $C_1 y_\lambda \ln^2 y_\lambda$ and the second by $C_2 y_\lambda \ln y_\lambda$, for some constants C_1 and C_2, so $|\zeta'(s)/\zeta(s)|$ is itself bounded by a constant multiple of $y_\lambda \ln^2 y_\lambda$. Hence

$$\left|\int_{\mathbf{BG}}\frac{\zeta'(s)}{\zeta(s)}\frac{x^s}{s^2}\,ds\right| \le C_3\frac{y_\lambda \ln^2 y_\lambda}{y_\lambda^2}\left|\int_{\mathbf{BG}}x^s\,ds\right| = C_4\frac{\ln^2 y_\lambda}{y_\lambda},$$

which approaches 0 as y_λ approaches ∞. Similar considerations show that the integral along **AH** also approaches 0 as y_λ approaches ∞.

Cauchy's integral theorem yields that

$$I_\Gamma = \frac{1}{2\pi i} \int_\Gamma \frac{\zeta'(s)}{\zeta(s)} \frac{x^s}{s^2}\, ds$$

is equal to the sum of the residues at the poles of $\zeta'(s)$ and the roots of $\zeta(s)$ that lie within Γ. The only such pole occurs at $s = 1$, where the residue is $-x$, and by construction, the sum of the residues of the roots inside Γ cannot exceed ϵx. Thus, writing $I_\Gamma = I_{\mathbf{AB}} + I_{\mathbf{BG}} + I_{\mathbf{GE}} + I_{\mathbf{ECDF}} + I_{\mathbf{FH}} + I_{\mathbf{HA}}$ and taking the limit as λ approaches ∞,

$$-x \le -\psi_2(x) + 0 + \lim_{\lambda\to\infty} I_{\mathbf{GE}} + I_{\mathbf{ECDF}} + \lim_{\lambda\to\infty} I_{\mathbf{FH}} + 0 \le -x + \epsilon x.$$

The quantity $I_{\mathbf{ECDF}}$ is a constant, since the boundary segment **ECDF** is fixed regardless of the value of λ. So $\lim_{x\to\infty}(I_{\mathbf{ECDF}}/x) = 0$, and to show that $\lim_{x\to\infty}(\psi_2(x)/x) = 1$, it remains only to show that

$$\lim_{x\to\infty} \frac{\lim_{\lambda\to\infty} I_{\mathbf{GE}}}{x} = \lim_{x\to\infty} \frac{\lim_{\lambda\to\infty} I_{\mathbf{FH}}}{x} = 0.$$

For that, Hadamard noted that for s on **GE** and **FH** (regardless of the value of λ, which determines the position of **G** and **H**) and any root ρ of $\zeta(s)$, the ratio $\left|\dfrac{s - \rho}{\rho}\right|$ must be greater than a fixed constant, so by (39), $\left|\dfrac{1}{s}\dfrac{\zeta'(s)}{\zeta(s)}\right|$ is finite. Therefore for some constant K, $I_{\mathbf{GE}}$ is less than $K \int_{GE} |x^s|/|s|\, ds$ — a finite quantity — and likewise for $I_{\mathbf{FH}}$. So $\dfrac{I_{\mathbf{GE}}}{x}$ and $\dfrac{I_{\mathbf{FH}}}{x}$ both approach 0 as x approaches ∞.

10.3   The proof of de la Vallée Poussin

In a note appended to the end of Hadamard (1896b), Hadamard remarked that while correcting the proofs for that paper he had been notified of de la Vallée Poussin's paper on the same topic (de la Vallée Poussin 1896), and he acknowledged that their proofs, found independently, had some points in common. In particular, both involved showing that $\zeta(s)$ had no roots on the line $\text{Re}\, s = 1$. But their methods for doing so were entirely different, and Hadamard judged that his own method was simpler.[7]

---

[7]In a statement quoted on page 198 of Narkiewicz (2000), de la Vallée Poussin agreed, but nevertheless claimed priority for the proof of that result.

Indeed, of the twenty-one pages in Hadamard (1896b), only two and a half are devoted to his proof that $\zeta(s)$ has no root of the form $1 + \beta i$; and as Hadamard noted, that proof rested on only two simple properties of $\zeta(s)$: that its logarithm could be expressed as a series of the form $\sum a_n e^{-\lambda_n s}$, for some positive constants a_n and λ_n, and that ζ itself was analytic on the limiting line of convergence of that series, with the exception of a single simple pole.

In contrast, de la Vallée Poussin's proof that $\zeta(s)$ had no roots on the line $\mathrm{Re}\, s = 1$ was the subject of the third chapter of his paper de la Vallée Poussin (1896), which took up 18 of its 74 pages. Like Hadamard's proof, de la Vallée Poussin's was by contradiction and used the fact that any root $s = 1 + \beta i$ of $\zeta(s)$ would have to be a simple root. But unlike Hadamard's argument, which rested on boundedness considerations, de la Vallée Poussin's was based on the uniqueness of certain Fourier series expansions. Specifically, he considered complex-valued functions $f(y)$ of a real variable y that for $y > 1$ have the form $L(y) + P(y)$, where $L(y)$ denotes a function that approaches a finite limit A as y approaches ∞ and $P(y)$ denotes an infinite series of the form

$$\sum_{n=1}^{\infty} [c_n \cos(\alpha_n \ln y) + d_n \sin(\alpha_n \ln y)],$$

in which the coefficients α_n do not approach zero as n approaches ∞ and the series $\sum c_n$ and $\sum d_n$ are absolutely convergent. He proved that in any such representation the values A, c_n, d_n and α_n are all uniquely determined by $f(y)$. On the other hand, if $1 + \beta i$ were a root of $\zeta(s)$, he showed by a long and delicate argument that the function $\dfrac{1 + \cos(\beta \ln y)}{y} \sum_{p<y} \ln p$, where p ranges over primes, would have two distinct representations of the form $L(y) + P(y)$. Salient details of the computations involved are given on pp. 208–214 of Narkiewicz (2000).

Two other differences between the tools used by Hadamard and those employed by de la Vallée Poussin are worth noting.

First, although de la Vallée Poussin referred to Riemann's functional equation for the ζ-function and observed that it could be used to define $\zeta(s)$ throughout $\mathbb{C} - \{1\}$, he did not make use of that extension in his proof of the Prime Number Theorem. Instead, he noted that integration by parts yields

$$\int_0^1 \frac{dx}{(n+x)^s} = \frac{1}{(n+1)^s} + s \int_0^1 \frac{x\,dx}{(n+x)^{s+1}},$$

from which it follows that for $\mathrm{Re}\, s > 1$,

$$\zeta(s) = 1 + \sum_{n=1}^{\infty} \left(\int_0^1 \frac{dx}{(n+x)^s} - s \int_0^1 \frac{x\,dx}{(n+x)^{s+1}} \right).$$

But

$$\sum_{n=1}^{\infty} \int_0^1 \frac{dx}{(n+x)^s} = \int_1^{\infty} \frac{dx}{x^s} = \frac{1}{s-1},$$

so

(40) $$\zeta(s) = 1 + \frac{1}{s-1} - s \sum_{1}^{\infty} \int_0^1 \frac{x\,dx}{(n+x)^{s+1}};$$

and since the sum on the right side of (40) converges absolutely for $\mathrm{Re}\,s > 0$, that equation may be used to define $\zeta(s)$ in that half-plane, except for a simple pole at $s = 1$ with residue 1.

Second, in place of the integral $\psi_2(x)$ employed by Hadamard, both in his proof that no $s = 1 + \beta i$ is a root of $\zeta(s)$ and in his proof that $\pi(x) \sim x/\ln x$ de la Vallée Poussin made use of the integral

$$I_{u,v}(y) = \frac{1}{2\pi i} \int_{a-\infty i}^{a+\infty i} \frac{\zeta'(s)}{\zeta(s)} \frac{y^s\,ds}{(s-u)(s-v)},$$

in which neither u nor v are poles or zeros of $\zeta(s)$, y is a real number greater than 1, and a is a real number greater than any of 1, $\mathrm{Re}\,u$ and $\mathrm{Re}\,v$.

Replacing the fraction $\zeta'(s)/\zeta(s)$ in the integrand of $I_{u,v}$ by the expression on the right side of (39) and integrating the result, de la Vallée Poussin obtained the identity

(41) $$I_{u,v}(y) = \frac{1}{u-v}\left(y^u \frac{\zeta'(u)}{\zeta(u)} - y^v \frac{\zeta'(v)}{\zeta(v)}\right) - \frac{y}{(u-1)(v-1)}$$

$$+ \sum_{\rho} \frac{y^\rho}{(u-\rho)(v-\rho)} + \sum_{m=1}^{\infty} \frac{y^{-2m}}{(2m+u)(2m+v)},$$

where ρ ranges over the roots of $\zeta(s)$, ordered according to increasing absolute value.

On the other hand, using (37) and (36), de la Vallée Poussin found that

(42) $$I_{u,v}(y) = -\frac{1}{u-v}\left(y^u \sum_{n<y} \frac{\Lambda(n)}{n^u} - y^v \sum_{n<y} \frac{\Lambda(n)}{n^v}\right),$$

valid for all u, v whenever $y > 1$.

Equating those two expressions for $I_{u,v}(y)$, solving for the quantity in parentheses in (42), setting $v = 0$, dividing by y^u and letting u approach 1 yields

(43) $$\sum_{n<y} \frac{\Lambda(n)}{n} - \frac{1}{y}\sum_{n<y} \Lambda(n) = \ln y - \lim_{u\to 1}\left(\frac{\zeta'(u)}{\zeta(u)} + \frac{u}{u-1}\right)$$

$$+ \frac{1}{y}\frac{\zeta'(0)}{\zeta(0)} - \sum_{\rho} \frac{y^{\rho-1}}{\rho(\rho-1)} - \sum_{m=1}^{\infty} \frac{y^{-2m-1}}{2m(2m+1)},$$

where the term $\ln y$ is obtained by writing $\frac{u}{1-u}y^{1-u}$ as $\frac{u}{1-u}[1 + (y^{1-u} - 1)]$ and applying l'Hopital's Rule to $\lim_{u\to 1} \frac{y^{1-u} - 1}{1-u}$.

Recalling the definition of $\Lambda(n)$, the left member of (43) can be written as
$$\sum_{p^m<y} \frac{\ln p}{p^m} - \frac{1}{y}\sum_{p^m<y}\ln p.$$ In the right member the fraction $\dfrac{u}{u-1}$ can be split up as $\dfrac{1}{u-1}+1$, and both summations there approach zero as y approaches ∞. (In particular, since the real part of any root ρ of $\zeta(s)$ has been shown to be less than one, the real part of $\rho-1$ must be negative. The terms of the first summation are thus dominated by those of the absolutely convergent series $\sum_\rho \dfrac{1}{\rho(\rho-1)}$, so that sum must converge *uniformly* to zero as y increases without bound.) Consequently, setting $u = s$, (43) takes the form

$$(44)\qquad \sum_{p^m<y}\frac{\ln p}{p^m} - \frac{1}{y}\sum_{p^m<y}\ln p = \ln y - 1 - \lim_{s\to 1}\left[\frac{\zeta'(s)}{\zeta(s)} + \frac{1}{s-1}\right] + f(y),$$

where $f(y)$ approaches 0 as y approaches ∞.

De la Vallée Poussin next noted that the difference $\sum_{p<y}\dfrac{\ln p}{p-1} - \sum_{p^m<y}\dfrac{\ln p}{p^m}$ approaches 0 as y approaches ∞, and proved that $\dfrac{1}{y}\sum_{p^m<y}\ln p - \dfrac{1}{y}\sum_{p<y}\ln p$ does so as well. He also proved that

$$\lim_{s\to 1}\left[\frac{\zeta'(s)}{\zeta(s)} + \frac{1}{s-1}\right] = \text{Euler's constant } \gamma,$$

defined as $\gamma = \lim_{n\to\infty}\left(\sum_{k=1}^{n} 1/k - \ln n\right)$.

Equation (44) can therefore be replaced by

$$(45)\qquad \sum_{p<y}\frac{\ln p}{p-1} - \frac{1}{y}\sum_{p<y}\ln p = \ln y - 1 - \gamma + g(y),$$

where $g(y)$ approaches 0 as y approaches ∞.

Equation (45) is the key to the final two steps in de la Vallée Poussin's derivation of the prime number theorem, namely

(i) showing that

$$(46)\qquad \int_1^x \frac{1}{y}\sum_{p<y}\ln p\, dy = x(1+h(x)),\ \text{where}\ \lim_{x\to\infty} h(x) = 0,$$

and

(ii) showing that (46) implies that $\lim_{x\to\infty}\dfrac{\theta(x)}{x} = 1$.

To establish (46) de la Vallée Poussin integrated (45) from 1 to x and then multiplied each term by $1/x$, thus obtaining

$$(47)\qquad \frac{1}{x}\int_1^x \sum_{p<y}\frac{\ln p}{p-1}\, dy - \frac{1}{x}\int_1^x \frac{1}{y}\sum_{p<y}\ln p\, dy = \frac{1}{x}\int_1^x \ln y\, dy - \gamma - 1 + j(x),$$

where $j(x)$ approaches 0 as x approaches ∞. He then noted that the first term in the left member of (47) can be rewritten as

(48) $$\frac{1}{x}\int_1^x \sum_{p<y} \frac{\ln p}{p-1}\, dy = \frac{1}{x}\sum_{p<x} \frac{\ln p}{p-1}\int_p^x dy$$

$$= \sum_{p<x} \frac{\ln p}{p-1} - \frac{1}{x}\sum_{p<x} \frac{p}{p-1}\ln p$$

$$= \sum_{p<x} \frac{\ln p}{p-1} - \frac{1}{x}\sum_{p<x} \ln p + \frac{1}{x}\sum_{p<x} \frac{\ln p}{p-1}.$$

The first two terms in the last member of (48) make up the left member of (45), with x in place of y. Moreover, as Franz Mertens had shown in 1874, it follows from Chebyshev's upper bound (31) for $\psi(x)$ that $\sum_{p<y} \dfrac{\ln p}{p-1}$ is less than a constant multiple of $\ln y$. Consequently,

$$\frac{1}{x}\int_1^x \sum_{p<y} \frac{\ln p}{p-1}\, dy = \ln x - \gamma - 1 + k(x),$$

where $k(x)$ approaches 0 as x approaches ∞. Substituting the expression on the right of this equation for the first integral in (47) and carrying out the integration in the right member of (47) then yields

$$\ln x - \gamma - 1 + k(x) - \frac{1}{x}\int_1^x \frac{1}{y}\sum_{p<y}\ln p\, dy = \frac{1}{x}[x\ln x - x] - \gamma - 1 + j(x),$$

which after cancellation of terms common to both members is (i), with $h(x) = k(x) - j(x)$.

To show that $\lim_{x\to\infty}\dfrac{\theta(x)}{x} = 1$, take $\epsilon > 0$, and use (46) to evaluate

(49) $$\int_x^{(1+\epsilon)x} \frac{1}{y}\sum_{p<y}\ln p\, dy = \int_1^{(1+\epsilon)x} \frac{1}{y}\sum_{p<y}\ln p\, dy - \int_1^x \frac{1}{y}\sum_{p<y}\ln p\, dy$$

$$= (1+\epsilon)x\,[1 + h((1+\epsilon)x)] - x[1 + h(x)]$$

$$= \epsilon x - h(x)x + (1+\epsilon)x h((1+\epsilon)x).$$

Dividing (49) by ϵx gives

$$\frac{1}{\epsilon x}\int_x^{(1+\epsilon)x} \frac{1}{y}\sum_{p<y}\ln p\, dy = 1 + \frac{(1+\epsilon)h((1+\epsilon)x) - h(x)}{\epsilon},$$

and since $\theta(y) = \sum_{p \le y} \ln p$, upper and lower bounds on the integrand give

$$\frac{\theta(x)}{x} \frac{\ln(1+\epsilon)}{\epsilon} \le 1 + \frac{(1+\epsilon)h((1+\epsilon)x) - h(x)}{\epsilon} \le \frac{\theta((1+\epsilon)x)}{x} \frac{\ln(1+\epsilon)}{\epsilon},$$

that is,

$$(50) \qquad \frac{\theta(x)}{x} \le \frac{\epsilon}{\ln(1+\epsilon)} + \frac{(1+\epsilon)h((1+\epsilon)x) - h(x)}{\ln(1+\epsilon)} \le \frac{\theta((1+\epsilon)x}{x}.$$

Since both $h(x)$ and $h((1+\epsilon)x)$ approach zero as x approaches ∞, (50) shows that

$$\limsup_{x \to \infty} \frac{\theta(x)}{x} \le \frac{\epsilon}{\ln(1+\epsilon)} \le \liminf_{x \to \infty} \frac{\theta((1+\epsilon)x)}{x};$$

and by replacing x in the latter inequality by $\dfrac{x}{1+\epsilon}$, it follows that

$$\frac{\epsilon}{\ln(1+\epsilon)} \le \liminf_{x \to \infty} \frac{\theta(x)}{x}(1+\epsilon).$$

By L'Hopital's rule, $\lim_{\epsilon \to 0} \dfrac{\epsilon}{\ln(1+\epsilon)} = 1$, so finally, $\lim_{x \to \infty} \dfrac{\theta(x)}{x} = 1$.

10.4   Later refinements

In the wake of Hadamard's and de la Vallée Poussin's proofs, various simplifications, generalizations, and improvements of their arguments were developed, by Edmund Landau, Franz Mertens, de la Vallée Poussin himself (who in 1899 obtained the error bound $\pi(x) - \mathrm{li}(x) \le Kxe^{-c\sqrt{\ln x}}$ for some positive constants K and c — a result not bettered for a quarter of a century thereafter[8]), and others. Modern proofs of the Prime Number Theorem that are descendants of the classical ones incorporate many of those refinements and also make use of other tools such as the Riemann-Lebesgue lemma, integral transforms, and Tauberian theorems (discussed further below).

The proof of the Prime Number Theorem given in Jameson (2003) may be taken as an exemplar of such proofs. Here, in outline, is its structure:

1. The comparison test is used to prove Dirichlet's result that if

$$\sum_{n=1}^{\infty} |a_n/n^\alpha|$$

---

[8]Cf. the discussion in Bateman and Diamond (1996), p. 736.

converges for some real α, then the corresponding Dirichlet series $\sum_{n=1}^{\infty} a_n/n^s$ converges absolutely for all $s = \sigma + it$ with $\sigma \geq \alpha$. In particular, if $|a_n| \leq 1$ for all n, then $\sum_{n=1}^{\infty} a_n/n^s$ converges absolutely when $\mathrm{Re}\, s > 1$, so $\sum_{n=1}^{\infty} 1/n^s$ can be used to define $\zeta(s)$ in that region.

2. It is shown that if $F(s) = \sum_{n=1}^{\infty} a_n/n^s$ converges whenever $\mathrm{Re}\, s > c$, then $F(s)$ is analytic at all points s with $\mathrm{Re}\, s > c$ and $F'(s) = -\sum_{n=1}^{\infty} a_n \ln n/n^s$. In particular, $\zeta'(s) = -\sum_{n=1}^{\infty} \ln n/n^s$ for $\mathrm{Re}\, s > 1$.

3. After defining the Möbius function $\mu(n)$ by $\mu(1) = 1$; $\mu(n) = 0$ if p^2 divides n for some prime p; and $\mu(n) = (-1)^k$ if $n = p_1 p_2 \ldots p_k$ for distinct primes $p_1, p_2, \ldots p_k$, the generalized Euler product identity (stated in footnote 3 above) is invoked and inverted to obtain

$$\frac{1}{\sum_{n=1}^{\infty} f(n)} = \prod_{p \, prime} (1 - f(p)) = \sum_{n=1}^{\infty} \mu(n) f(n).$$

Consequently $\dfrac{1}{\zeta(s)} = \sum_{n=1}^{\infty} \dfrac{\mu(n)}{n^s}$ for $\mathrm{Re}\, s > 1$.

4. It is verified that for $|s| < 1$ the series $\sum_{n=1}^{\infty} s^n/n$ defines an analytic function $h(s)$ that is a logarithm of $\dfrac{1}{1-s}$. The Euler product identity then implies that the logarithm of $\zeta(s)$ should be given by $H(s) = \sum_{p \, prime} h(p^{-s}) = \sum_{p \, prime} \sum_{n=1}^{\infty} \dfrac{1}{np^{ns}}$. It is proved that that double series converges when $\mathrm{Re}\, s > 1$ and that its sum is equal to that of the series $\sum_{n=1}^{\infty} \dfrac{c(n)}{n^s}$, where $c(n) = 1/m$ if $n = p^m$ for some prime p, and $c(n) = 0$ otherwise.

5. Consequently, $H'(s) = \dfrac{\zeta'(s)}{\zeta(s)} = -\sum_{n=1}^{\infty} \dfrac{c(n) \ln n}{n^s} = -\sum_{n=1}^{\infty} \dfrac{\Lambda(n)}{n^s}$, where $\Lambda(n)$ is the von Mangoldt function.

6. Abel summation is used to show that for any sequence $a(n)$ and corresponding summation function $A(x) = \sum_{n \leq x} a(n)$, if $X > 1$ then

$$\sum_{n \leq X} \frac{a(n)}{n^s} = \frac{A(X)}{X^s} + s \int_1^X \frac{A(x)}{x^{s+1}} \, dx.$$

Furthermore, if $s \neq 0$, $A(x)/x^s$ approaches 0 as x approaches ∞ and the Dirichlet series $\sum_{n=1}^{\infty} \dfrac{a(n)}{n^s}$ converges, then the Dirichlet integral

$$s \int_1^{\infty} \frac{A(x)}{x^{s+1}} \, dx$$

converges to the same value. Since $\psi(x)$ is the summation function for $\Lambda(n)$ and $\psi(x)/x^s$ approaches 0 as x approaches ∞ if $\mathrm{Re}\, s > 1$, it follows from 5 that

. (51)
$$\frac{\zeta'(s)}{\zeta(s)} = -s \int_1^\infty \frac{\psi(x)}{x^{s+1}} \, dx.$$

7. A simplification due to Edmund Landau is used to extend the domain of definition of $\zeta(s)$ without using the functional equation for $\zeta(s)$.[9] Specifically, for $\operatorname{Re} s > 0$, $\zeta(s)$ may be defined by

(52)
$$\zeta(s) = \frac{1}{s-1} + 1 - s \int_1^\infty \frac{x - [x]}{x^{s+1}} \, dx,$$

where $[x]$ denotes the greatest integer not exceeding x, and differentiation under the integral sign shows that $\zeta(s) - \dfrac{1}{s-1}$ is analytic at $s = 1$, so $\zeta(s)$ has a simple pole there. Furthermore,

$$\lim_{s \to 1} \left(\zeta(s) - \frac{1}{s-1} \right) = 1 - \int_1^\infty \frac{x - [x]}{x^2} \, dx = \text{Euler's constant } \gamma.$$

8. Hence $\zeta(s)$, $\dfrac{1}{\zeta(s)}$ and $\dfrac{\zeta'(s)}{\zeta(s)}$ are represented by Laurent series of the forms

$$\zeta(s) = \frac{1}{s-1} + \gamma + \sum_{n=1}^\infty c_n (s-1)^n$$

$$\frac{1}{\zeta(s)} = (s-1) - \gamma(s-1)^2 + \dots \quad \text{and}$$

$$\frac{\zeta'(s)}{\zeta(s)} = -\frac{1}{s-1} + a_0 + a_1(s-1) + \dots,$$

all converging in some disk with center $s = 1$.

9. Since

$$\left| \int_N^\infty \frac{x - [x]}{x^{s+1}} \, dx \right| \leq \int_N^\infty \frac{1}{x^{\sigma+1}} \, dx = \frac{1}{\sigma N^\sigma},$$

Euler's summation formula for finite sums[10] yields

$$\zeta(s) = \sum_{n=1}^N \frac{1}{n^s} + \frac{N^{1-s}}{s-1} - s \int_N^\infty \frac{x - [x]}{x^{s+1}} \, dx \leq \sum_{n=1}^N \frac{1}{n^s} + \frac{N^{1-s}}{s-1} + \frac{|s|}{\sigma N^\sigma}.$$

Straightforward calculations using the last inequality then show that when $\sigma \geq 1$ and $t \geq 2$, $|\zeta(\sigma + it)| \leq \ln t + 4$ and $|\zeta'(\sigma + it)| \leq \dfrac{1}{2}(\ln t + 3)^2$.

---

[9] According to Bateman and Diamond (1996), p. 737, Landau was the first to prove the PNT without recourse to that functional equation.

[10] $\sum_{n=2}^N f(n) = \int_1^N f(x) \, dx + \int_1^N (x - [x]) f'(x) \, dx$.

Those inequalities can in turn be used to show that $1/|\zeta(\sigma + it)| \leq 4(\ln t + 5)^7$ for $\sigma > 1$ and $t \geq 2$. (For details see Jameson 2003, pp. 108–109.)

10. The proof that $\zeta(s) \neq 0$ when $\mathrm{Re}\, s = 1$ is carried out by a simplified argument based on Hadamard's approach and later ideas of de la Vallée Poussin and Franz Mertens. It rests on the trigonometric identity $0 \leq 2(1 + \cos\theta)^2 = 3 + 4\cos\theta + \cos(2\theta)$. For if a Dirichlet series $\sum_{n=1}^{\infty} \dfrac{a(n)}{n^s}$ with positive real coefficients $a(n)$ converges for $\sigma > \sigma_0$ to a function $f(s)$, then for such σ

$$3f(\sigma) + 4f(\sigma + it) + f(\sigma + 2it) = \sum_{n=1}^{\infty} \frac{a(n)}{n^\sigma}(3 + 4n^{-it} + n^{-2it})$$

has real part

$$\sum_{n=1}^{\infty} \frac{a(n)}{n^\sigma}\mathrm{Re}(3 + 4n^{-it} + n^{-2it}) = \sum_{n=1}^{\infty} \frac{a(n)}{n^\sigma}(3 + 4\cos(t\ln n) + \cos(2t\ln n)) \geq 0.$$

In particular, by 4. the Dirchlet series for $\ln(\zeta(s))$ has positive coefficients and converges when $\sigma > 1$, so for such σ and all t,

$$3\ln(\zeta(\sigma)) + 4\mathrm{Re}\,\zeta(\sigma + it) + \mathrm{Re}\,\zeta(\sigma + 2it)$$

$$= \ln\left[\zeta(\sigma)^3|\zeta(\sigma + it)|^4|\zeta(\sigma + 2it)|\right] \geq 0$$

(because $\mathrm{Re}\,\ln z = \ln|z|$), that is

(53) $$\zeta(\sigma)^3|\zeta(\sigma + it)|^4|\zeta(\sigma + 2it)| \geq 1.$$

Suppose then that $\zeta(1 + it_0) = 0$ for some $t_0 \neq 0$. Then

(54) $$\zeta(\sigma)^3|\zeta(\sigma + it_0)|^4|\zeta(\sigma + 2it_0)|$$

$$= [(\sigma - 1)\zeta(\sigma)]^3 \left(\frac{|\zeta(\sigma + it_0)|}{\sigma - 1}\right)^4 (\sigma - 1)|\zeta(\sigma + 2it_0)|.$$

But as σ approaches 1^+, $(\sigma - 1)\zeta(\sigma)$ approaches 1, while $\dfrac{\zeta(\sigma + it_0)}{\sigma - 1}$ approaches $\zeta'(1 + it_0)$ and $\zeta(\sigma + 2it_0)$ approaches $\zeta(1 + 2it_0)$. So by (54), the product $\zeta(\sigma)^3|\zeta(\sigma + it_0)|^4|\zeta(\sigma + 2it_0)|$ approaches 0, contrary to (53).

11. Cauchy's integral theorem is used to evaluate three infinite contour integrals, namely

(55) $$\frac{1}{2\pi i}\int_{c-i\infty}^{c+i\infty} \frac{x^s}{s^2}\,ds = S(x)\ln x \qquad \text{for } x > 0 \text{ and } c > 0$$

(56) $$\frac{1}{2\pi i}\int_{c-i\infty}^{c+i\infty} \frac{x^s}{s(s-1)}\,ds = (x - 1)S(x) \qquad \text{for } x > 0 \text{ and } c > 1$$

Fig. 10.2 Contour used to
evaluate the integrals (53)
through (55). Adapted from
The Prime Number Theorem,
p. 116, by G.J.O. Jameson.
Reprinted with the
permission of Cambridge
University Press

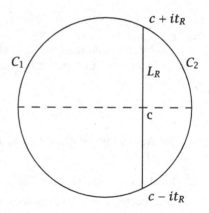

(57)

$$\frac{1}{2\pi i} \int_{c-i\infty}^{c+i\infty} \frac{x^{s-1}}{s(s-1)} f(s)\, ds = \int_1^x \frac{A(y)}{y^2}\, dy \quad \text{for } x > 1 \text{ and } c > 1,$$

where $S(x)$ denotes the step-function equal to 0 for $x < 1$ and 1 for $x \geq 1$
and $A(x) = \sum_{n \leq x} a(n)$ is the summation function corresponding to a Dirichlet
series $\sum_{n=1}^{\infty} a(n)/n^s$ that converges absolutely to $f(s)$ whenever $\mathrm{Re}\, s > 1$.

In all three cases the infinite contour integrals are obtained as the limit,
as R approaches ∞, of integrals taken along the paths $\Gamma_1 = C_1 \cup L_R$ or
$\Gamma_2 = C_2 \cup L_R$, where C_1 and C_2 are the arcs of the circle of radius R centered
at $s = 0$ that lie, respectively, to the left and right of the vertical line segment
L_R with endpoints $c - t_R$ and $c + t_R$ on the circle. (See Figure 10.2.)

The integrand of (55) may be rewritten as

$$\frac{x^s}{s^2} = \frac{e^{ls}}{s^2} = \frac{1}{s^2} \sum_{n=0}^{\infty} \frac{l^n s^n}{n!},$$

where $l = \ln x$. Since the exponential series converges uniformly on any closed
interval, the integration may then be carried out term by term. For $x \geq 1$ the
path Γ_1 is taken, enclosing the pole at $s = 0$ of x^s/s^2, where the residue is l.
For $0 < x < 1$ the path Γ_2 is taken instead, which encloses no poles of x^s/s^2;
so that integral is 0. In the first case, $|s| = R$ and $|x^s| = x^\sigma \leq x^c$ on C_1, so the
integral along C_1 has absolute value less than or equal to $\dfrac{1}{2\pi} \dfrac{x^c}{R^2} 2\pi R = \dfrac{x^c}{R}$,
which approaches 0 as R approaches ∞. In the second case, $x^\sigma \leq x^c$ on C_2 as
well, since for $0 < x < 1$, x^σ decreases as σ increases. The integral along C_2
thus also approaches 0 as R approaches ∞. The proof of (56) is similar, using
$|s(s-1)| \geq R(R-1)$ and writing $\dfrac{x^s}{s(s-1)}$ as $\dfrac{x^s}{s-1} - \dfrac{x^s}{s}$.

To obtain (57), first note that $x^s f(s) = \sum_{n \leq x} a(n) \left(\dfrac{x}{n}\right)^s + \sum_{n > x} a(n) \left(\dfrac{x}{n}\right)^s$.
The first sum is finite, and so can be integrated term by term. By (56), the
result is $\sum_{n \leq x} S\left(\dfrac{x}{n}\right) \left(\dfrac{x}{n} - 1\right) = \sum_{n \leq x} a(n) \left(\dfrac{x}{n} - 1\right)$. The second term,

on the other hand, is an analytic function of s that for $\text{Re}\, s \geq c$ is bounded by $\sum_{n>x} |a(n)|(x/n)^c$. By Cauchy's theorem, its integral over Γ_2 is zero, and by the same reasoning as before, its integral over C_2 tends to zero as R approaches ∞. Consequently, after dividing by x,

$$\frac{1}{2\pi i} \int_{c-i\infty}^{c+i\infty} \frac{x^{s-1}}{s(s-1)} f(s)\, ds = \sum_{n \leq x} a(n) \left(\frac{1}{n} - \frac{1}{x} \right).$$

By Abel summation applied to the function $1/y$, the last sum is equal to $\int_1^x A(y)/y^2\, dy$.

12. Finally, the Prime Number Theorem in the form $\lim_{x\to\infty} \dfrac{\psi(x)}{x} = 1$ is obtained as a special case of the following much more general result.

Theorem: *Suppose the function $f(s)$ is analytic throughout the region $\text{Re}\, s \geq 1$, except perhaps at $s = 1$, and satisfies the following conditions:*

(C1) $f(s) = \sum_{n=1}^{\infty} \dfrac{a(n)}{n^s}$ *converges absolutely when $\text{Re}\, s > 1$.*

(C2) $f(s) = \dfrac{\alpha}{s-1} + \alpha_0 + (s-1)h(s)$, *where h is differentiable at $s = 1$.*

(C3) *There is a function $P(t)$ such that $|f(\sigma+it)| \leq P(t)$ for $\sigma \geq 1$ and $t \geq t_0 \geq 1$, and $\int_1^{\infty} P(t)/t^2\, dt$ is convergent.*

Then $\displaystyle\int_1^{\infty} \frac{A(x) - \alpha x}{x^2}\, dx$ *converges to $\alpha_0 - \alpha$ where $A(x) = \sum_{n \leq x} a(n)$.*

If, furthermore, $A(x)$ is increasing and non-negative, then $\lim_{x\to\infty} \dfrac{A(x)}{x} = \alpha$.

To prove the first claim, note that $\dfrac{1}{s-1} = \dfrac{s}{s-1} - 1$, so $(s-1)h(s) = f(s) - \alpha\dfrac{s}{s-1} - (\alpha_0 - \alpha)$, and $|(s-1)h(s)| \leq P(t) + |\alpha| + |\alpha_0|$ when $|\sigma \geq 1$ and $|t| \geq t_0$. Then for $x > 1$ and $c > 1$, (55), (56), and (57) give

$$\frac{1}{2\pi i} \int_{c-i\infty}^{c+i\infty} \frac{x^{s-1}h(s)}{s}\, ds = \frac{1}{2\pi i} \int_{c-i\infty}^{c+i\infty} \frac{x^{s-1}}{s(s-1)} f(s)\, ds$$

$$- \frac{\alpha}{2\pi i} \int_{c-i\infty}^{c+i\infty} \frac{x^{s-1}}{(s-1)^2}\, ds - \frac{\alpha_0 - \alpha}{2\pi i} \int_{c-i\infty}^{c+i\infty} \frac{x^{s-1}}{s(s-1)}\, ds$$

$$= \int_1^x \frac{A(y)}{y^2}\, dy - \alpha \ln x - (\alpha_0 - \alpha)\left(1 - \frac{1}{x}\right)$$

$$= \int_1^x \frac{A(y) - \alpha y}{y^2}\, dy - (\alpha_0 - \alpha)\left(1 - \frac{1}{x}\right).$$

Careful examination (see Jameson 2003, p. 121) shows that the result just obtained also holds for $c = 1$. That is,

$$\frac{1}{2\pi i} \int_{1-i\infty}^{1+i\infty} \frac{x^{s-1}h(s)}{s}\, ds = \int_1^x \frac{A(y) - \alpha y}{y^2}\, dy - (\alpha_0 - \alpha)\left(1 - \frac{1}{x}\right).$$

The path of integration for the contour integral on the left is the vertical line $\mathrm{Re}\, s = 1$, where $s = 1 + it$, so that integral may be rewritten as

$$\frac{1}{2\pi}\int_{-\infty}^{+\infty}\frac{x^{it}h(1+it)}{1+it}\,dt = \frac{1}{2\pi}\int_{-\infty}^{+\infty}\frac{e^{it\ln x}h(1+it)}{1+it}\,dt$$

$\left(\text{which converges absolutely since } \left|\dfrac{h(s)}{s}\right| \le \dfrac{P(t)+|\alpha|+|\alpha_0|}{t^2}\right).$ That the latter
integral approaches 0 as x approaches ∞ then follows from the Riemann-Lebesgue lemma, which states that

If $\phi(t)$ is a continuously differentiable function from $\mathbb{R}$ to $\mathbb{C}$ and $\int_{-\infty}^{+\infty}|\phi(t)|\,dt$ converges, then the integral

$$\int_{-\infty}^{+\infty} e^{i\lambda t}\phi(t)\,dt$$

approaches 0 as λ approaches ∞.

The convergence of $\int_1^\infty (A(x)-\alpha x)/x^2\,dx$ implies the remaining claim (that if $A(x)$ is increasing and non-negative, then $\lim_{x\to\infty}\dfrac{A(x)}{x} = \alpha$). For such convergence means that for any $\delta > 0$, there is an R such that

$$\left|\int_{x_0}^{x_1}\frac{A(x)-x}{x^2}\,dx\right| < \delta \quad \text{whenever} \quad x_1 > x_0 > R.$$

Then for $0 < \delta < 1/2$, the assumption either that $A(x_0) > (1+\delta)x_0$ for some $x_0 > R$ (and hence for all $x \ge x_0$, since $A(x)$ is increasing), or that $A(x_0) < (1-\delta)x_0$ for some $x_0 \ge 2R$, leads to a contradiction. (See Jameson 2003, p. 131.)

The Prime Number Theorem follows from the Theorem by taking $f(s) = -\dfrac{\zeta'(s)}{\zeta(s)}$: For the result of step 5. shows that $f(s) = \sum_{n=1}^{\infty}\dfrac{\Lambda(n)}{n^s}$ (whence (C1) is satisfied); $\psi(x) = \sum_{n\le x}\Lambda(n)$ is increasing; (C2) holds by the third equation in step 8., with $\alpha = 1$; and the inequalities found in step 9. show that $|f(s)| = |f(\sigma+it)| < P(t) = 2(\ln t + 5)^9$. Since $\int_1^\infty P(t)/t^2\,dt$ converges, (C3) is thus also satisfied.

10.5 Tauberian theorems and Newman's proof

The Theorem stated in the previous section is an example of a so-called Tauberian theorem, broadly defined (as in Edwards 1974, p. 279) as a theorem that "permits a conclusion about one kind of average [in this case, $A(x)/x$] given information about another kind of average [here, the integral from 1 to ∞ of $(A(x)-\alpha x)/x^2$]."

The Prime Number Theorem was first deduced from a Tauberian theorem by Edmund Landau in his paper Landau (1908).[11] Seven years later Hardy and Littlewood gave another such proof (Hardy and Littlewood 1915), and in the 1930s Tauberian theorems based on Norbert Wiener's methods in the theory of Fourier transforms were employed to deduce the Prime Number Theorem from the non-vanishing of $\zeta(s)$ on the line $\mathrm{Re}\, s = 1$. In particular, the Prime Number Theorem was so deduced using

The Wiener-Ikehara Theorem: Suppose that f is a non-decreasing real-valued function on $[1, \infty)$ for which $\int_1^\infty |f(u)|u^{-\sigma-1}\, du$ converges for all real $\sigma > 1$. If in addition $\int_1^\infty f(u)u^{-s-1}\, du = \alpha/(s-1) + g(s)$ for some real α, where $g(s)$ is the restriction to $\mathrm{Re}\, s > 1$ of a function that is continuous on $\mathrm{Re}\, s \geq 1$, then $\lim_{u\to\infty} f(u)/u = \alpha$.

But proofs of the Wiener-Ikehara Theorem are themselves difficult.

In 1935 A.E. Ingham proved another Tauberian theorem that related the Laplace transform $\int_0^\infty f(t)e^{-zt}\, dt$ of a function $f(t)$ defined on $[0, \infty)$ to the integral of $f(t)$ itself over that interval.[12] But that proof, too, was complicated, and was also based on results from Fourier analysis (Ingham 1935). In 1980, however, D.J. Newman found a way to prove a variant of Ingham's theorem and to derive the Prime Number Theorem from it without resort either to Fourier techniques or to contour integrals over infinite paths.

Newman's original proof (Newman 1980) was couched in terms of Dirichlet series: he proved that if $\sum_{n=1}^\infty a_n n^{-s}$ converges to an analytic function $f(s)$ for all s with $\mathrm{Re}\, s > 1$, if $|a_n| \leq 1$ for every n, and if $f(s)$ is also analytic when $\mathrm{Re}\, s = 1$, then $\sum_{n=1}^\infty a_n n^{-s}$ converges for all s with $\mathrm{Re}\, s = 1$. Subsequent refinements of that proof, as given in Korevaar (1982), Zagier (1997), chapter 7 of Lax and Zalcman (2012), and Jameson (2003), recast it as an alternative proof of Ingham's Tauberian theorem. The formulation of that result given in Lax and Zalcman (2012) reads:

Let f be a bounded measurable function on $[0, \infty)$. Suppose that the Laplace transform

$$g(z) = \int_0^\infty f(t)e^{-zt}\, dt,$$

which is defined and analytic on the open half plane $\{z : \mathrm{Re}\, z > 0\}$, extends analytically to an open set containing $\{z : \mathrm{Re}\, z \geq 0\}$. Then the improper integral $\int_0^\infty f(t)\, dt = \lim_{T\to\infty} \int_0^T f(t)\, dt$ converges and coincides with $g(0)$, the value of the analytic extension of g at $z = 0$.

---

[11]Discussed in detail in Narkiewicz (2000), pp. 298–302.

[12]Ingham's theorem may alternatively be stated in terms of the Mellin transform $\int_1^\infty f(t)t^{-s}\, dt$. See, e.g., Korevaar (1982) or Jameson (2003), pp. 124–129.

Fig. 10.3 Contours used in
Newman's proof

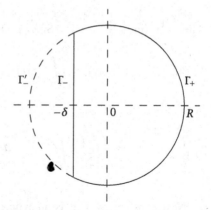

The proof there proceeds as follows:

Say $|f(t)| \le M$ for all $t \ge 0$. For $T > 0$ let $g_T(z)$ be the entire function defined by $\int_0^T f(t)e^{-zt}\,dt$. The theorem then asserts that $\lim_{T \to \infty}|g(0) - g_T(0)| = 0$.

To establish that, choose $R > 0$ and $\delta(R)$ sufficiently small that g is analytic throughout the region $D = \{z : |z| \le R \text{ and } \operatorname{Re} z \ge -\delta(R)\}$. Let Γ be the boundary of D (shown as the solid curve in Figure 10.3), traversed counterclockwise, and consider

$$(58) \qquad \frac{1}{2\pi i}\int_\Gamma [g(z) - g_T(z)]e^{zT}\left(1 + \frac{z^2}{R^2}\right)\frac{1}{z}\,dz.$$

By Cauchy's theorem, the value of (58) is $g(0) - g_T(0)$. Writing $\sigma = \operatorname{Re} z$, if $\sigma > 0$ then

$$(59) \qquad |g(z) - g_T(z)| = \left|\int_T^\infty f(t)e^{-zt}\,dt\right| \le M\int_T^\infty e^{-\sigma t}\,dt = \frac{Me^{-\sigma T}}{\sigma}.$$

Also, when $|z| = R$,

$$(60) \qquad \left|e^{zT}\left(1 + \frac{z^2}{R^2}\right)\frac{1}{z}\right| = e^{\sigma T}\left|\left(\frac{1}{z} + \frac{z}{|z|^2}\right)\right| = e^{\sigma T}\left|\frac{1}{z} + \frac{1}{\bar{z}}\right| = e^{\sigma T}\frac{2|\sigma|}{R^2}.$$

Now let Γ_+ be the semicircle $\Gamma \cap \{\operatorname{Re} z > 0\}$, let Γ_- denote $\Gamma \cap \{\operatorname{Re} z < 0\}$, and let Γ'_- be the semicircle $\{z : |z| = R \text{ and } \operatorname{Re} z < 0\}$. By (59) and (60), for $z \in \Gamma_+$, the absolute value of the integrand in (58) is bounded by $2M/R^2$, so

$$(61) \qquad \left|\frac{1}{2\pi i}\int_{\Gamma_+}[g(z) - g_T(z)]e^{zT}\left(1 + \frac{z^2}{R^2}\right)\frac{1}{z}\,dz\right| \le \frac{M}{R}.$$

For $z \in \Gamma_-$, first consider g_T. As an entire function, its integral over Γ_- is the same as its integral over Γ'_-. Since for $\sigma < 0$,

$$(62) \qquad |g_T(z)| = \left| \int_0^T f(t) e^{-zt} \, dt \right| \leq M \int_{-\infty}^T e^{-\sigma t} \, dt = \frac{M e^{-\sigma T}}{|\sigma|},$$

it follows from (60) that

$$(63) \qquad \left| \frac{1}{2\pi i} \int_{\Gamma'_-} g_T(z) e^{zT} \left(1 + \frac{z^2}{R^2} \right) \frac{1}{z} \, dz \right| \leq \frac{M}{R}.$$

Next consider $g(z)$. It is analytic on Γ_-, so the quantity $\left| g(z) \left(1 + \dfrac{z^2}{R^2} \right) \dfrac{1}{z} \right|$ is bounded on Γ_- by some constant K (whose value depends on δ and R). Likewise, e^{zT} is bounded on Γ_-, and converges uniformly to 0 on compact subsets of $\{\operatorname{Re} z < 0\}$ as T approaches ∞. Consequently,

$$(64) \qquad \lim_{T \to \infty} \left| \frac{1}{2\pi i} \int_{\Gamma_-} g(z) e^{zT} \left(1 + \frac{z^2}{R^2} \right) \frac{1}{z} \, dz \right| = 0.$$

Recalling that $g(0) - g_T(0)$ is given by the integral in (58), it follows from (61), (63), and (64) that

$$\limsup_{T \to \infty} |g(0) - g_T(0)| \leq \frac{2M}{R}.$$

Since that holds for arbitrarily large values of R, $\lim_{T \to \infty} |g(0) - g_T(0)| = 0$.

The Prime Number Theorem in the form $\lim_{x \to \infty} \psi(x)/x = 1$ is then deduced, as in the previous section, from the convergence of the improper integral $\int_1^\infty [\psi(x) - x]/x^2 \, dx$. That, in turn, follows from Ingham's Tauberian theorem by setting $x = e^t$ and taking $f(t) = \psi(e^t) e^{-t} - 1$. For, by equation (51) above,

$$-\frac{\zeta'(s)}{\zeta(s)} = s \int_1^\infty \frac{\psi(x)}{x^{s+1}} \, dx = s \int_0^\infty \psi(e^t) e^{-st} \, dt,$$

so

$$g(s) = \int_0^\infty f(t) e^{-st} \, dt = \int_0^\infty [\psi(e^t) e^{-t} - 1] e^{-st} \, dt$$

$$= \int_0^\infty \psi(e^t) e^{-(s+1)t} \, dt - \int_0^\infty e^{-st} \, dt$$

$$= \frac{1}{s+1} \left[-\frac{\zeta'(s+1)}{\zeta(s+1)} \right] - \frac{1}{s} = \frac{1}{s+1} \left[-\frac{\zeta'(s+1)}{\zeta(s+1)} - \frac{1}{s} - 1 \right].$$

Chebyshev's upper bound for $\psi(x)$ shows that $f(t)$ is bounded, and the Laurent series for $\zeta'(s)/\zeta(s)$ given in step 8 of the previous section shows that the expression in the last member of the equation above can be extended to an analytic function on the half plane $\mathrm{Re}\, s \geq 0$. Hence the hypotheses of Ingham's theorem are satisfied.

10.6 Elementary proofs

In 1909 Edmund Landau published an influential handbook (Landau 1909) that "presented in accessible form nearly everything that was then known about the distribution of prime numbers" (Bateman and Diamond 1996, p. 737). It popularized use of the O-notation[13] in statements concerning growth rates of functions, and drew attention to the power of complex-analytic methods in number theory.

In particular, in contrast to the elementary methods of Chebyshev, those of complex analysis had yielded the Prime Number Theorem. The question thus arose: Were such methods *essential* to the proof of that theorem?

Many leading number theorists came to believe that they were. G.H. Hardy, for example, in an address to the Mathematical Society of Copenhagen in 1921, declared

> No elementary proof of the prime number theorem is known, and one may ask whether it is reasonable to expect one. [For] ... we know that ... theorem is roughly equivalent to ... the theorem that Riemann's zeta function has no roots on a certain line.[14] A proof of such a theorem, not fundamentally dependent on the theory of functions, [thus] seems to me extraordinarily unlikely."   (Quoted from Goldfeld 2004.)

In 1948, however, Atle Selberg and Paul Erdős, independently but each using results of the other, found ways to prove the Prime Number Theorem without reference to the ζ-function or complex variables and without resort to methods of Fourier analysis. Their proofs, however, were 'elementary' only in that technical sense. Indeed, in Edwards (1974) Harold Edwards expresses the widely shared opinion that "Since 1949 many variations, extensions and refinements of [Selberg's and Erdős's] elementary proof[s] have been given, but none of them seems very straightforward or natural, nor does any of them give much insight into the theorem." Furthermore, as noted in Jameson (2003), p. 207, no elementary proof so far devised has given error estimates for the approximation of $\pi(x)$ by $\mathrm{li}(x)$] that are "as strong

---

[13]Whereby $f(x) = O(g(x))$ for $x > x_1 \geq x_0$ means that f is eventually dominated by g, that is, that f and g are both defined for $x > x_0$, $g(x) > 0$ for $x > x_0$, and there is a constant K such that $|f(x)| \leq Kg(x)$ for all $x > x_1$.

[14]For as noted in the preceding section, the Wiener-Ikehara Theorem implies that the Prime Number Theorem follows from the absence of zeroes of the ζ-function on the line $\mathrm{Re}\, s = 1$, a fact that is also implied by the Prime Number Theorem. (See, for example, Diamond 1982, pp. 572–573.)

as that of de la Vallée Poussin." The interest in such proofs would thus seem to stem primarily from concern for purity of method.

The basis of most elementary proofs of the Prime Number Theorem is a growth estimate for a logarithmic summation found by Selberg. In one form it states that for $x > 1$

$$(65) \qquad \sum_{p \leq x} (\ln p)^2 + \sum_{pq \leq x} \ln p \ln q = 2x \ln x + O(x),$$

where p and q denote prime numbers. Other statements equivalent to (65), shown to be so in Diamond (1982), p. 566 and Jameson (2003), pp. 214–215, are

$$(66) \qquad \theta(x) \ln x + \sum_{p \leq x} \theta(x/p) \ln p = 2x \ln x + O(x)$$

$$(67) \qquad \sum_{n \leq x} [\Lambda(n) \ln n + (\Lambda * \Lambda)(n)] = 2x \ln x + O(x)$$

$$(68) \qquad \psi(x) \ln x + \sum_{n \leq x} \Lambda(n) \psi(x/n) = 2x \ln x + O(x) \quad \text{and}$$

$$(69) \qquad R(x) \ln x + \sum_{n \leq x} \Lambda(n) R(x/n) = O(x) \quad (\text{where } R(x) = \psi(x) - x),$$

in which the symbol $*$ denotes the Dirichlet convolution operation on arithmetic sequences, defined by $a * b = \sum_{jk=n} a(j)b(k) = \sum_{j|n} a(j)b(n/j)$.

Selberg's original proof derived the equation $\lim_{x \to \infty} \theta(x)/x = 1$ from (66) using a consequence of (66) discovered by Erdős. In particular, denoting $\liminf \theta(x)/x$ by a and $\limsup \theta(x)/x$ by A, Selberg deduced from (66) that $a + A = 2$. Meanwhile, unaware of that fact, Erdős used (66) to show that for any $\delta > 0$ there is a constant $K(\delta)$ such that for sufficiently large values of x there are more than $K(\delta)x/\ln x$ primes in the interval $(x, x + K(\delta)x)$. Erdős communicated his proof of that fact to Selberg, who then, via an intricate argument, used Erdős's result to prove that $A \leq a$. Consequently, $A = a = 1$. (See Erdős 1949 for details of all those proofs. The proof given in Selberg 1949 is a later, more direct one that does not use Erdős's result.[15])

The most accessible elementary proof of the Prime Number Theorem is probably that in Levinson (1969), whose very title ("A motivated elementary proof of the Prime Number Theorem") suggests that the strategies underlying elementary proofs

---

[15] Regrettably, the interaction between Erdős and Selberg in this matter was a source of lasting bitterness between them. Goldfeld (2004) provides a balanced account of the dispute, based on primary sources. As noted there, the issue was not one of priority of discovery, but "arose over the question of whether a joint paper (on the entire proof) or separate papers (on each individual contribution) should appear".

of that theorem are not perspicuous.[16] Variants of Levinson's proof are also given in
Edwards (1974) and Jameson (2003). Here, in outline, is the structure of the latter
version:

1. The goal is to show that $\lim_{x\to\infty} \psi(x)/x = 1$. Toward that end, three related
   functions whose behavior is easier to study are defined, namely

$$R(x) = \begin{cases} 0 & \text{if } x < 1 \\ \psi(x) - x & \text{if } x \geq 1 \end{cases}, \quad S(x) = \int_0^x \frac{R(t)}{t}\, dt \quad \text{and} \quad W(x) = \frac{S(e^x)}{e^x}.$$

   Then $\lim_{x\to\infty} \psi(x)/x = 1$ if and only if $\lim_{x\to\infty} R(x)/x = 0$.

2. It follows from Chebyshev's result $\psi(x) \leq 2x$ that $|R(x)| \leq x$ for $x > 0$.
   Furthermore, $\int_1^x R(t)/t^2\, dt = \int_1^x \psi(t)/t^2\, dt - \ln x$, and Abel summation yields

$$\int_1^x \frac{\psi(t)}{t^2}\, dt = \sum_{n \leq x} \frac{\Lambda(n)}{n} - \frac{\psi(x)}{x}.$$

   On the other hand, Mertens in 1874 applied Chebyshev's bound to obtain that
   $\sum_{n \leq x} \Lambda(n)/n = \ln x + O(1)$. (See Jameson 2003, p. 90 for details.) Consequently, applying Chebyshev's bound once more, $\int_1^x R(t)/t^2\, dt$ is bounded for
   $x > 1$.

3. Since, by 2., the absolute value of the integrand in the definition of S is
   bounded by 1, S satisfies the Lipschitz condition $|S(x_2) - S(x_1)| \leq x_2 - x_1$
   for $x_2 > x_1 > 0$. That, in turn, together with the inequality $e^{-x} \geq 1 - x$, shows
   that W likewise satisfies the Lipschitz condition $|W(x_2) - W(x_1)| \leq 2(x_2 - x_1)$
   for $x_2 > x_1 > 0$.

4. The Lipschitz condition on S gives $|S(x)| \leq x$ for $x > 0$, that is, $|S(x)/x| \leq 1$
   for $x > 0$. Then $|W(x)| \leq 1$ and

$$\int_1^x \frac{S(t)}{t^2}\, dt = \int_1^x \frac{1}{t^2} \int_1^t \frac{R(u)}{u}\, du\, dt$$

$$= \int_1^x \frac{R(u)}{u} \int_u^x \frac{1}{t^2}\, dt\, du$$

$$= \int_1^x \frac{R(u)}{u} \left(\frac{1}{u} - \frac{1}{x} \right) du$$

$$= \int_1^x \frac{R(u)}{u^2}\, du - \frac{S(x)}{x},$$

---

[16]Levinson's paper won the Mathematical Association of America's Chauvenet Prize for exposition in 1971. Nevertheless, after reading it, the number theorist Harold Stark commented "Well,
Norman tried, but the thing is as mysterious as ever." (Quoted in Segal 2009, p. 99.)

which is bounded for all $x > 1$ by the result of step 2. above. Consequently, $\int_0^x W(t)\, dt$ is bounded for all $x > 0$.

5. Straightforward arguments with inequalities yield the following Tauberian theorem:

If $A(x) \geq 0$, $A(x)$ is increasing for $x > 1$, and $\dfrac{1}{x} \displaystyle\int_1^x \dfrac{A(t)}{t}\, dt \to 1$ as $x \to \infty$,

then $\dfrac{A(x)}{x} \to 1$ as $x \to \infty$.

Then since $S(x)/x = (1/x) \int_1^x \psi(t)/t\, dt - 1 - 1/x$, taking $A(x) = \psi(x)$ in the Tauberian theorem shows that to prove that $\psi(x)/x \to 1$ as $x \to \infty$ it suffices to prove that $S(x)/x \to 0$ as $x \to \infty$.

6. Equivalently, it suffices to show that $W(x) \to 0$ as $x \to \infty$. For that purpose, let

$$\alpha = \limsup_{x \to \infty} |W(x)| \leq 1 \quad \text{and} \quad \beta = \limsup_{x \to \infty} \frac{1}{x} \int_0^x |W(t)|\, dt.$$

Then $\beta \leq \alpha$. Crucially, however,

(70) $\beta = \alpha \quad$ only if $\quad \alpha = 0;$

so to prove the Prime Number Theorem it suffices to show that $\alpha \leq \beta$.

To prove (70), assume $\alpha > 0$. Since $\int_0^x W(t)\, dt$ is bounded for all $x > 0$, there is a constant B such that $|\int_{x_1}^{x_2} W(t)\, dt| \leq B$ for all $x_2 > x_1 > 0$. Also, by the definition of α, for any $a > \alpha$ there is some x_a such that $|W(x)| \leq a$ for $x > x_a$. So suppose $\alpha < a \leq 2\alpha$ and consider $\int_{x_1}^{x_1+h} |W(t)|\, dt$, where $x_1 \geq x_a$ and $h \geq 2\alpha$ is to be determined. If $W(x)$ changes sign within the interval $[x_1, x_1 + h]$, the intermediate-value theorem yields the existence of a point z in that interval where $W(z) = 0$. The Lipschitz condition on W then gives $|W(x)| \leq 2|x - z|$, and since $h \geq a$, at least one of the points $z \pm \dfrac{a}{2}$ lies between x_1 and $x_1 + h$. So either the interval $[z, z + a/2]$ lies within $[x_1, x_1 + h]$ or $[z - a/2, z]$ does. Whichever does, call it I. (If both do, pick one.) Then $\int_I |W(x)|\, dx \leq \int_I 2|x - z|\, dx = a^2/4$. The part of $[x_1, x_1 + h]$ lying outside I has length $h - a/2$, and there $|W(x)| \leq a$. So, finally,

(71) $\displaystyle\int_{x_1}^{x_1+h} |W(x)|\, dx \leq \frac{a^2}{4} + a\left(h - \frac{a}{2}\right) = a\left(h - \frac{a}{4}\right) < a\left(h - \frac{\alpha}{4}\right).$

By choosing h to be the greater of α and $B/\alpha + \alpha/4$, (71) can be ensured to hold as well if $W(x)$ does not change sign within $[x_1, x_1 + h]$.

To complete the proof, note that for any $x \geq x_1 + h$ there is an integer n such that $x_a + nh \leq x < x_a + (n + 1)h$. Then

$$\int_0^x |W(x)|\,dx \le \int_0^{x_a} |W(x)|\,dx + (n+1)\left(h - \frac{\alpha}{4}\right)ah$$

$$= C + (n+1)\left(h - \frac{\alpha}{4}\right)ah,$$

where C is constant and $n \to \infty$ as x does. Hence, since $x > nh$,

$$\frac{1}{x}\int_0^x |W(x)|\,dx \le \frac{C}{x} + \left(1 + \frac{1}{n}\right)\left(1 - \frac{\alpha}{4h}\right)a.$$

As $x \to \infty$, the right member of that inequality approaches $(1 - \alpha/4h)a$. Consequently $\beta \le (1-\alpha/4h)a$ for *any* $a > \alpha$. Therefore $\beta \le (1-\alpha/4h)\alpha < \alpha$.

7. It is at this point that Selberg's inequality enters in. In Levinson (1969) and Jameson (2003) that inequality, in the forms (68) and (69), is derived as a corollary to the *Tatuzawa-Iseki identity*, which states that if F is a function defined on the interval $[1, \infty)$ and $G(x) = \sum_{n\le x} F(x/n)$, then for $x \ge 1$:

$$(72) \qquad \sum_{k\le x}\mu(k)\ln\frac{x}{k}G\left(\frac{x}{k}\right) = F(x)\ln x + \sum_{n\le x}\Lambda(n)F\left(\frac{x}{n}\right),$$

where μ denotes the Möbius function and Λ the von Mangoldt function.

The Tatuzawa-Iseki identity is a variant of the *Möbius inversion formula*, which states that under the same hypotheses, $F(x) = \sum_{n\le x}\mu(n)G(x/n)$. It is obtained by multiplying the Möbius formula by $\ln x$ to get the first term in the right member of (72), rewriting the factor $\ln(x/k)$ in the left member as $\ln x - \ln k$, replacing $G\left(\frac{x}{k}\right)$ in $\sum_{k\le x}\mu(k)\ln k\, G\left(\frac{x}{k}\right)$ by $\sum_{j\le x/k} F\left(\frac{x}{jk}\right)$, interchanging the order of summation in the double sum, and expressing the result in terms of $\Lambda(n)$.[17]

To obtain inequality (68), the Tatuzawa-Iseki identity is applied with $F(x) = R(x) + \gamma + 1$, where γ is Euler's constant. One proves that then $|G(x)| \le \ln x + 2$ for $x \ge 1$. The derivation is completed by invoking the integral test for series together with Chebyshev's upperbound for $\psi(x)$.

8. Inequality (67) is deduced as a corollary to (68), using Abel summation and Chebyshev's result that $\psi(x) \le 2x$. Since the integral test implies that $\sum_{n\le x}\ln n = x\ln x + O(x)$, a further equivalent of Selberg's formula is $\sum_{n\le x}[\Lambda(n)\ln n + (\Lambda * \Lambda)(n)] - 2\ln n = O(x)$.

9. Dividing (69) by t, where $1 \le t \le x$, gives

$$R(t)\frac{\ln t}{t} + \frac{1}{t}\sum_{n\le x}\Lambda(n)R\left(\frac{t}{n}\right) = O(1).$$

---

[17] The Möbius inversion formula by itself does not suffice to give the desired bound on $R(x)$, and that, in Levinson's opinion, accounts for "the long delay in the discovery of an elementary proof" of the Prime Number Theorem. (Levinson 1969, p. 235)

Integrating and using $|S(t)/t| \le 1$ from step 4. then shows that for $x \ge 1$,

$$(73) \qquad S(x)\ln x + \sum_{n \le x} \Lambda(n) S(x/n) = O(x),$$

which is (69) with R replaced by S.

10. Three technical lemmas are now proved, using (68), (69) and the Lipschitz condition on S, respectively. First, for all $x \ge 1$ there is a constant K_1 such that

$$(74) \quad |S(x)|(\ln x)^2 \le \sum_{n \le x} [\Lambda(n)\ln n + (\Lambda * \Lambda)(n)] \left| S\left(\frac{x}{n}\right) \right| + K_1 x \ln x.$$

Second,
$$(75)$$
$$\sum_{n \le x} [\Lambda(n)\ln n + (\Lambda * \Lambda)(n)] \left| S\left(\frac{x}{n}\right) \right| = 2 \sum_{n \le x} \ln n \left| S\left(\frac{x}{n}\right) \right| + O(x \ln x);$$

and third, for all $x \ge 1$ there is a constant K_2 such that

$$(76) \qquad \sum_{n \le x} \ln n \left| S\left(\frac{x}{n}\right) \right| \le \int_1^x \ln t \left| S\left(\frac{x}{t}\right) \right| dt + K_2 x.$$

11. Together, (74), (75), and (76) yield that for all $x \ge 1$ there is a constant K_3 such that

$$|S(x)|(\ln x)^2 \le 2 \int_1^x \ln t \left| S\left(\frac{x}{t}\right) \right| dt + K_3 x \ln x, \qquad \text{so that}$$

$$(77) \qquad |W(x)| \le \frac{2}{x^2} \int_0^x (x-u)|W(u)| \, du + \frac{K_3}{x}.$$

12. Finally, for α and β as defined in step 6., (77) implies that $\alpha \le \beta$. For, by the definition of β, for every $\epsilon > 0$ there exists an x_1 such that for every $x \ge x_1$, $\int_0^x |W(t)| \, dt \le (\beta + \epsilon)x$. In order to apply (77), consider $\int_0^x (x-u)|W(u)| \, du$, which may be rewritten as

$$\int_0^x |W(u)| \int_u^x dt \, du = \int_0^x \int_0^t |W(u)| \, du \, dt.$$

Then for $x \ge x_1$,

$$\int_{x_1}^x \int_0^t |W(u)| \, du \, dt \le \int_{x_1}^x (\beta + \epsilon)t \, dt = \frac{1}{2}(\beta + \epsilon)(x^2 - x_1^2).$$

On the other hand, $|W(u)| \leq 1$ for all u by step 4. So

$$\int_0^{x_1} \int_0^t |W(u)| \, du \, dt \leq \int_0^{x_1} t \, dt = \frac{1}{2} x_1^2.$$

Therefore

$$\int_0^x (x - u)|W(u)| \, du = \int_0^{x_1} (x - u)|W(u)| \, du + \int_{x_1}^x (x - u)|W(u)| \, du$$

$$\leq \frac{1}{2}(\beta + \epsilon)x^2 + \frac{1}{2} x_1^2.$$

Then by (77), $|W(x)| \leq \beta + \epsilon + \dfrac{x_1^2}{x^2} + \dfrac{K_3}{x}$. By definition of α, that means $\alpha \leq \beta + \epsilon$; and since ϵ can be chosen to be arbitrarily small, $\alpha \leq \beta$.   q.e.d.

10.7 Overview

The five proofs of the Prime Number Theorem considered here employ a wide range of methodologies, including analytic continuation, Abel summation, Dirichlet convolution, contour integration, Fourier analysis, Laplace transforms, and Tauberian theorems. They differ from one another in many respects, including the ways in which the domain of definition of $\zeta(s)$ is extended and whether or not recourse is made to the functional equation for the ζ-function.

The proofs also exemplify several of the different motivations discussed in Chapter 2. For example, Riemann's program for proving the Prime Number Theorem by examining the behavior of the complex ζ-function proposed bringing the methods of complex analysis to bear on the seemingly remote field of number theory; so the successful carrying out of that program by Hadamard and de la Vallée Poussin may be deemed instances of benchmarking. And the marked differences in the independent and nearly simultaneous proofs of the theorem that Hadamard and de la Vallée Poussin gave — each of which was based upon the same corpus of earlier work (especially Chebyshev's results) and followed the same basic steps in Riemann's program (proving that $\zeta(s)$ has no roots of the form $1 + it$ and then applying complex contour integration to expressions involving ζ'/ζ) — are surely attributable to differences in their individual patterns of thought.

In the case of Hadamard, a further motive was that of correcting deficiencies in Cahen's earlier proof attempt. Indeed, at the beginning of section 12 of his memoir, Hadamard noted explicitly that Cahen had claimed to have proved that $\theta(x) \sim x$, but that his demonstration was based on an unsubstantiated claim by Stieltjes (who thought he had proved the Riemann Hypothesis). Nonetheless, Hadamard declared, "we will show that the same result can be obtained in a completely rigorous way" via "an easy modification" of Cahen's analysis.[18]

The modern descendants of those classical proofs illustrate how simplifications, generalizations, and refinements of mathematical arguments gradually evolve and are incorporated into later proofs. Examples include the proof that $\zeta(s)$ has no roots with real part equal to one, which was significantly shortened and simplified through the use of a trigonometric identity; the deduction of the Prime Number Theorem from a general Tauberian theorem, which showed it to be a particular instance of a family of such theorems concerning Dirichlet series;[19] and the use in Newman's proof of a contour integral that is much more easily evaluated than the classical ones. Newman's proof also exhibits economy of means with regard to conceptual prerequisites, making it comprehensible to those having only a rudimentary knowledge of complex analysis.

The 'elementary' proofs, on the other hand, are no simpler than the classical analytic ones, are generally regarded as *less* perspicuous (giving little or no insight into *why* the Prime Number Theorem is true), and do not yield as sharp error bounds as those obtained by analytic means. The esteem nonetheless accorded them by the mathematical community, as reflected in the prizes awarded to their discoverers, may thus be taken as a quintessential manifestation of the high regard mathematicians have for purity of method in and of itself.

References

Bateman, P., Diamond, H.: A hundred years of prime numbers. Amer. Math. Monthly **103**, 729–741 (1996)

Chebyshev, P.: Sur la fonction qui détermine la totalité des nombres premiers inférieurs à une limite donnée. Mémoires des savants étrangers de l'Acad. Sci. St.Pétersbourg **6**, 1–19 (1848)

Chebyshev, P.: Mémoire sur nombres premiers. Mémoires des savants étrangers de l'Acad. Sci. St.Pétersbourg **7**, 17–33 (1850)

de la Vallée Poussin, C.: Recherches analytiques sur la théorie des nombres premiers. Ann. Soc. Sci. Bruxelles **20**, 183–256 (1896)

Diamond, H.: Elementary methods in the study of the distribution of prime numbers. Bull. Amer. Math. Soc. **7**, 553–589 (1982)

Edwards, H.M.: Riemann's Zeta Function. Academic Press, New York and London (1974)

Erdős, P.: On a new method in elementary number theory which leads to an elementary proof of the prime number theorem. Proc. Nat. Acad. Sci. U.S.A. **35**, 374–384 (1949)

[18]"Nous allons voir qu'en modifiant légèrement l'analyse de l'auteur on peut établir le même résultat en toute rigueur."

[19]Three other instances given in Jameson (2003) are $\sum_{n=1}^{\infty} \frac{\mu(n)}{n} = 0$, where μ denotes the Möbius function, $\sum_{n=1}^{\infty} \frac{\mu(n)}{n^{1+it}} = \frac{1}{\zeta(1+it)}$, and $\sum_{n=1}^{\infty} \frac{(-1)^{\Omega(n)}}{n} = 0$, where $\Omega(n)$ denotes the number of prime factors of n, each counted according to its multiplicity.

Goldfeld, D.: The elementary proof of the prime number theorem: an historical perspective. Number Theory (New York Seminar, 2003). Springer, New York, 179–192 (2004)

Goldstein, L.J.: A history of the prime number theorem. Amer. Math. Monthly **80**, 599–615 (1973)

Hadamard, J.: Étude sur les propriétés des fonctions entières et en particulier d'une fonction considérée par Riemann. J. math. pures appl. (4) **9**, 171–215 (1893)

Hadamard, J.: Sur les zéros de la fonction $\zeta(s)$ de Riemann. Comptes Rendus Acad. Sci. Paris **122**, 1470–1473 (1896a)

Hadamard, J.: Sur la distribution des zéros de la fonction $\zeta(s)$ et ses conséquences arithmétiques. Bull. Soc. Math. France **24**, 199–220 (1896b)

Hardy, G.H., Littlewood, J.: New proofs of the prime number theorem and similar theorems. Quart. J. Math., Oxford ser. **46**, 215–219 (1915)

Ingham, A.E.: On Wiener's method in Tauberian theorems. Proc. London Math. Soc. (2)**38**, 458–480 (1935)

Jameson, G.J.O.: The Prime Number Theorem. London Mathematical Society Student Texts **53**. Cambridge U.P., Cambridge (2003)

Korevaar, J.: On Newman's quick way to the Prime Number Theorem. Math. Intelligencer **4:3**, 108–115.

Landau, E.: Zwei neue Herleitungen für die asymptotische Anzahl der Primzahlen unter einer gegebener Grenze. SBer. Kgl. Preuß. Akad. Wiss. Berlin, 746–764 (1908)

Landau, E.: Handbuch der Lehre von der Verteilung der Primzahlen. Teubner, Leipzig (1909)

Lax, P.D., Zalcman, L.: Complex Proofs of Real Theorems. University Lecture Series 58. Amer. Math. Soc., Providence (2012)

Levinson, N.: A motivated account of an elementary proof of the prime number theorem. Amer. Math. Monthly **76**, 225–245 (1969)

Narkiewicz, W.: The Development of Prime Number Theory. Springer, Berlin (2000)

Newman, D.J.: Simple analytic proof of the prime number theorem. Amer. Math. Monthly **87**, 693–696 (1980)

Riemann, B.: Ueber die Anzahl der Primzahlen unter einer gegebener Größe. Monatsberichte der Königlichen Preußische Akademie der Wissenschaften zu Berlin, 671–680 (1860)

Segal, J.: Recountings: Conversations with MIT Mathematicians. A K Peters, Natick, Mass. (2009)

Selberg, A.: An elementary proof of the prime number theorem. Ann. Math. (2) **50**, 305–313 (1949)

von Mangoldt, H.: Zu Riemanns Abhandlung "Ueber die Anzanl der Primzahlen unter einer gegebenen Größe". J. reine u. angewandte Math. **114**, 255–305 (1895)

Zagier, D.: Newman's short proof of the Prime Number Theorem. Amer. Math. Monthly **104**, 705–708 (1997)

Chapter 11
The Irreducibility of the Cyclotomic Polynomials

by
Steven H. Weintraub

The irreducibility of the cyclotomic polynomials is a fundamental result in algebraic number theory that has been proved many times, by many different authors, in varying degrees of generality and using a variety of approaches and methods of proof. We examine these in the spirit of our inquiry here.

The cyclotomic polynomials were considered by Gauss, in the prime case, in the seventh section of his *Disquisitiones Mathematicae* (Gauss 1801). This section culminates in his famous result that a regular 17-gon, or, more generally, a regular p-gon, where p is a Fermat prime (i.e., a prime of the form $2^{2^m} + 1$), is constructible by ruler and compass. But he begins by considering these polynomials for an arbitrary prime p. (Throughout this chapter p will denote a prime.) He does not name these polynomials or introduce a notation for them, but we will use what is now standard mathematical terminology and notation in our discussion.

We let $\zeta_p = \exp(2\pi i/p) = \cos(2\pi/p) + i \sin(2\pi/p)$. Then for any positive integer k, $\zeta_p^k = \exp(2k\pi i/p) = \cos(2k\pi/p) + i \sin(2k\pi/p)$. In particular, ζ_p^k, $k = 0, \ldots, p-1$, are the p complex p-th roots of 1. (Of course, $\zeta_p^0 = 1$.) In other words, they are all the complex roots of the polynomial $x^p - 1$, whence $x^p - 1 = (x-1)(x-\zeta_p)\cdots(x-\zeta_p^{p-1})$ and so $(x^p-1)/(x-1) = (x-\zeta_p)\cdots(x-\zeta_p^{p-1})$. By definition this is the p-th *cyclotomic polynomial*

$$\Phi_p(x) = \frac{x^p - 1}{x - 1} = \prod_{k=1}^{p-1} (x - \zeta_p^k).$$

It is easy to write down $\Phi_p(x)$ explicitly. By elementary long division of polynomials

$$\Phi_p(x) = x^{p-1} + x^{p-2} + \cdots + x + 1.$$

Gauss proves (*Disquisitiones*, article 341):

Theorem: *For any prime p, the polynomial $\Phi_p(x)$ is irreducible, i.e., it is not the product of two polynomials of lower degree with rational coefficients.*

Let us remark now that in the course of his proof, Gauss uses a simplification of the problem that is used by all his successors as well.

© Springer International Publishing Switzerland 2015
J.W. Dawson, Jr., *Why Prove it Again?*, DOI 10.1007/978-3-319-17368-9_11

Gauss had previously proved the following result, now universally known as Gauss's Lemma (*Disquisitiones*, article 42):

Lemma: *Let $f(x)$ be a monic polynomial, i.e., a polynomial such that the coefficient of the highest power of x is 1, with integer coefficients. If $f(x) = g(x)h(x)$ with $g(x)$ and $h(x)$ monic polynomials with rational coefficients, then $g(x)$ and $h(x)$ have integer coefficients.*

This lemma immediately implies that a monic polynomial $f(x)$ is irreducible over the rationals if and only if it is irreducible over the integers (since if $f(x) = g_0(x)h_0(x)$ with $g_0(x)$ and $h_0(x)$ having rational coefficients, one could multiply each of $g_0(x)$ and $h_0(x)$ by constants so that they become monic). Thus it suffices to show that $\Phi_p(x)$ is not the product of two monic polynomials of lower degree with integer coefficients.

We next see how to generalize the definition of the cyclotomic polynomials. To do so, for an arbitrary positive integer n, we define ζ to be a *primitive n-th root of 1* if n is the smallest positive integer such that $\zeta^n = 1$. Letting $\zeta_n = \exp(2\pi i/n)$, it is easy to see that the primitive n-th roots of 1 are the complex numbers ζ_n^k, where k takes on all values between 0 and $n - 1$ that are relatively prime to n. Then we define the n-th cyclotomic polynomial by

$$\Phi_n(x) = \prod_k{}' (x - \zeta_n^k)$$

where the prime denotes that the product is taken over just those values of k.

In case $n = p$ is prime, those are all values between 1 and $n - 1$, but for n composite that is not the case. In general, there are $\varphi(n)$ such values of k, where $\varphi(n)$ is the Euler totient function, so $\Phi_n(x)$ is a polynomial of degree $\varphi(n)$.

If $n = p^r$ is a prime power, then an n-th root of 1 is either a primitive n-th root of 1 or a (not necessarily primitive) p^{r-1}-th root of 1, so in this case

$$\Phi_{p^r}(x) = \frac{x^{p^r} - 1}{x^{p^{r-1}} - 1} = x^{p^{r-1}(p-1)} + x^{p^{r-1}(p-2)} + \cdots + x^{p^{r-1}} + 1,$$

but for an arbitrary composite value of n, there is no simple formula for $\Phi_n(x)$.

However, there is an inductive way of finding $\Phi_n(x)$. Observe that if g is the gcd of k and n, then ζ_n^k is a primitive n/g-th root of 1. Thus, letting $d = n/g$, and observing that as g runs through the divisors of n, so does d, we see that every n-th root of 1 is a primitive d-th root of 1 for some d dividing n. Once again, the polynomial $x^n - 1$ has as its roots all of the n-th roots of 1, so

$$x^n - 1 = \prod_{k=0}^{n-1} (x - \zeta_n^k).$$

Thus

$$x^n - 1 = \prod_{d \text{ dividing } n} \Phi_d(x)$$

and so

$$\Phi_n(x) = \frac{x^n - 1}{\prod_{d \text{ a proper divisor of } n} \Phi_d(x)}.$$

We then inductively see, from Gauss's Lemma, that $\Phi_n(x)$ is a monic polynomial with integer coefficients, for every positive integer n.

We make the following elementary observation, used without comment through-out many of these proofs: If ζ is any n-th root of 1 other than 1 itself, then

$$1 + \zeta + \zeta^2 + \cdots + \zeta^{n-1} = \frac{\zeta^n - 1}{\zeta - 1} = 0$$

and of course, if $\zeta = 1$, then

$$1 + \zeta + \zeta^2 + \cdots + \zeta^{n-1} = n.$$

The general theorem in this regard is then:

Theorem: *For any positive integer n, the polynomial $\Phi_n(x)$ is irreducible, i.e., it is not the product of two polynomials of lower degree with rational coefficients.*

Furthermore, again by Gauss's Lemma, we observe that to prove this it suffices to prove that $\Phi_n(x)$ is not the product of two monic polynomials of lower degree with integer coefficients.

A second thread that runs through many of the proofs of this theorem is the use of the Fundamental Theorem on Symmetric Polynomials.

Consider a monic polynomial $f(x)$ of degree m,

$$f(x) = x^m + a_{m-1}x^{m-1} + \cdots + a_1 x + a_0.$$

Let this polynomial have roots $r_1, \ldots, r_m$. Then

$$f(x) = (x - r_1)(x - r_2) \cdots (x - r_m).$$

Then, expanding this polynomial, we see

$$x^m + a_{m-1}x^{m-1} + \cdots + a_1 x + a_0 = x^m - s_1(r_1, \ldots, r_m)x^{m-1}$$
$$+ s_2(r_1, \ldots, r_m)x^{m-2} - \cdots + (-1)^m s_m(r_1, \ldots, r_m)$$

where $s_k(r_1, \ldots, r_m)$ is the k-th *elementary symmetric function* in the roots $r_1, \ldots, r_m$, that is, the sum of the k-fold products of distinct roots. (Thus $s_1(r_1, \ldots, r_m) = r_1 + \cdots + r_m$, $s_2(r_1, \ldots, r_m) = r_1 r_2 + \cdots + r_{m-1} r_m$, $\cdots$, $s_m(r_1, \ldots, r_m) = r_1 \cdots r_m$.) We see that the coefficients of $f(x)$ are up to sign the values of the elementary symmetric polynomials in its roots; to be precise, $a_{m-i} = (-1)^i s_i(r_1, \ldots, r_m)$.

We call a polynomial in m variables *symmetric* if it is invariant under any permutation of the variables. The second result we need is the Fundamental Theorem on Symmetric Polynomials, which was first explicitly written down and proved by Lagrange, but which certainly goes back much farther.

Theorem: *Let $f(x)$ be a monic polynomial with rational (respectively, integer) coefficients. Let $g(x)$ be any symmetric polynomial in the roots of $f(x)$ with rational (respectively, integer) coefficients. Then $g(x)$ can be written as a polynomial in the elementary symmetric polynomials of the roots of $f(x)$, and hence as a polynomial in the coefficients of $f(x)$, with rational (respectively, integer) coefficients.*

We now present Gauss's proof that $\Phi_p(x)$ is irreducible (Gauss 1801, article 341).

Gauss remarks that the result is trivial for $p = 2$ (since $\Phi_2(x) = x + 1$), so we may assume p is odd.

Gauss begins with the following observation, which we shall single out as a lemma. (Actually, he only observes it in the rational case, as he will use Gauss's Lemma to obtain the conclusion in the integral case for the polynomial he is considering.)

Lemma: *Let $g(x)$ be an arbitrary polynomial with rational (respectively, integer) coefficients. Let $g(x)$ have roots $r_1, \ldots, r_m$. For a positive integer k, set $g_k(x) = (x - r_1^k) \cdots (x - r_m^k)$ (so that the roots of $g_k(x)$ are the k-th powers of the roots of $g(x)$). Then $g_k(x)$ has rational (respectively, integer) coefficients.*

To prove this, Gauss observes that the coefficients of $g_k(x)$ are symmetric functions in $r_1, \ldots, r_m$, so that this follows immediately from the Fundamental Theorem on Symmetric Polynomials.

As the second step in his proof, Gauss makes the following observation. Let $\varphi(x_1, x_2, x_3, \ldots)$ be any polynomial with integer coefficients. Let ζ be any primitive p-th root of 1. Then for any $k_1, k_2, k_3, \ldots$, writing the value

$$r_1 = \varphi(\zeta^{k_1}, \zeta^{k_2}, \zeta^{k_3}, \ldots) = A_0 + A_1 \zeta + A_2 \zeta^2 + \cdots + A_{p-1} \zeta^{p-1}$$

then for any integer t,

$$r_t = \varphi(\zeta^{tk_1}, \zeta^{tk_2}, \zeta^{tk_3}, \ldots) = A_0 + A_1 \zeta^t + A_2 \zeta^{2t} + \cdots + A_{p-1} \zeta^{(p-1)t}$$

and also, in particular,

$$r_p = \varphi(\zeta^{pk_1}, \zeta^{pk_2}, \zeta^{pk_3}, \ldots) = \varphi(1, 1, 1, \ldots) = A_0 + A_1 + A_2 + \cdots + A_{p-1}$$

and so the sum $r_1 + r_2 + \cdots + r_p = pA_0$ is divisible by p.

Now suppose that $\Phi_p(x) = f(x)g(x)$ for nonconstant monic polynomials $f(x)$ and $g(x)$ with integer coefficients. Write

$$f(x) = x^d + a_{d-1}x^{d-1} + \cdots + a_1 x + a_0.$$

Then $f(x)$ has d distinct roots, $g(x)$ has $(p-1)-d$ distinct roots, and they have no roots in common (as every root of $\Phi_p(x)$ must be a root of one of these polynomials, and the roots $\zeta_p, \zeta_p^2, \ldots, \zeta_p^{(p-1)}$ of $\Phi_p(x)$ are all distinct).

Let F be the set of roots of $f(x)$ and G be the set of roots of $g(x)$. (Of course, we are using anachronistic language here.) Let F' be the set of reciprocals of elements of F, all of which are also primitive p-th roots of 1, and similarly for G'. Note that for any primitive p-th root ζ of 1, its reciprocal is its complex conjugate. Also, since p is odd, $\zeta^{-1} = \bar{\zeta} \neq \zeta$.

We let $f'(x)$ be the monic polynomial whose roots are the elements of F', and observe that $f'(x) = x^d + (a_1/a_0)x^{d-1} + \cdots + (a_{d-1}/a_0)x + (1/a_0)$.

Gauss distinguishes four cases.

Case 1: $F = F'$. Then the roots of $f(x)$ occur in conjugate pairs, so $f(x)$ is a product of $d/2$ factors each of the form

$$(x-\zeta)(x-\bar{\zeta}) = x^2-(\zeta+\bar{\zeta})x+1 = x^2-2x\cos\theta+1 = (x-\cos\theta)^2+\sin^2\theta > 0$$

for every real number x, where $\zeta = \cos\theta + i\sin\theta$. Then $q_1 = f(1)$ is a positive integer. Set $f_1(x) = f(x)$ and let $f_k(x)$ be the monic polynomial whose roots are the k-th powers of the roots of $f(x)$, $k = 1, \ldots, p-1$. By the same argument, $q_k = f_k(1)$ is a positive integer for each k. Also, $f_p(x) = (x-1)^d$ so $q_p = f_p(1) = 0$.

Denote the elements of F by $\zeta_1, \ldots, \zeta_d$. (Here the subscripts denote different primitive p-th roots of 1.) Let $\varphi(x_1, \ldots, x_d)$ be the polynomial

$$\varphi(x_1, \ldots, x_d) = (1 - x_1) \cdots (1 - x_d).$$

Then $q_k = \varphi(\zeta_1^k, \ldots, \zeta_d^k)$, so by our previous observation

$$q_1 + \cdots + q_{p-1} = q_1 + \cdots + q_{p-1} + q_p \quad \text{is divisible by } p.$$

But also

$$f_1(x) \cdots f_{p-1}(x) = \Phi_p(x)^d$$

as every primitive p-th root of 1 is a root of $f_1(x) \cdots f_{p-1}(x)$ of multiplicity d. Hence, setting $x = 1$, we have

$$q_1 \cdots q_{p-1} = p^d.$$

Since p is a prime, $d < p - 1$, and each q_i is a positive integer, we must have g of the integers $q_1, \ldots, q_{p-1}$ equal to 1 and the rest divisible by p, for some $g > 0$, so

$$q_1 + \cdots + q_{p-1} \equiv g \not\equiv 0 \ (\text{mod } p),$$

a contradiction.[1]

Case 2: $F \neq F'$ but $T = F \cap F' \neq \emptyset$. Let $t(x)$ be the monic polynomial whose roots are the elements of T. Then $t(x)$ is the greatest common divisor of $f(x)$ and $f'(x)$. By the argument for case 1, $t(x)$ cannot have rational coefficients. But $f(x)$ and $g(x)$ each have rational coefficients, hence so does their greatest common divisor; contradiction.

Case 3: $G \cap G' \neq \emptyset$. Applying the argument of case 1 or case 2 to the polynomial $g(x)$ yields a contradiction.

Case 4: $G = F'$ and $F = G'$. Then every primitive p-th root of 1 is a root of $f(x)$ or of $f'(x)$, so $\Phi_p(x) = f(x) f'(x)$. Setting $x = 1$ in the expressions for $f(x)$ and $f'(x)$, and multipying through by a_0, we obtain

$$a_0 p = (1 + a_{d-1} + \cdots + a_0)^2.$$

But $f'(x)$ divides $\Phi_p(x)$, so has integer coefficients, and hence $a_0 = \pm 1$. This gives that $\pm p$ is a perfect square, which is impossible.   q.e.d.

There are two things to note about Gauss's proof. First, it applies only to the polynomial $\Phi_p(x)$. Second, it is quite involved.

There matters stood for some time. But beginning in the 1840s, there was rapid progress.

The next step was due to Kronecker (Kronecker 1845). As with Gauss, his proof only applies to the polynomial $\Phi_p(x)$. But his explicit motivation was to give a simpler proof of that important result.

Kronecker begins with the following lemma.

Lemma: *Let $f(x)$ be an arbitrary polynomial with integer coefficients. Let ζ be any primitive p-th root of 1. Then $f(\zeta) f(\zeta^2) \cdots f(\zeta^{p-1})$ is an integer and $f(\zeta) \cdots f(\zeta^{p-1}) \equiv f(1)^{p-1} \ (\text{mod } p)$.*

---

[1]Since $f(x)$ is a polynomial with rational coefficients, if ζ is a root of $f(x)$, i.e., $0 = f(\zeta)$, then $0 = \overline{0} = \overline{f(\zeta)} = f(\overline{\zeta}) = f(\zeta^{-1})$, i.e., ζ^{-1} is also a root of $f(x)$. Thus Case 1 is the only case that actually occurs. But Gauss does not make that observation.

Proof: Observe that

$$f(\zeta)f(\zeta^2)\cdots f(\zeta^{p-1})$$

is a symmetric polynomial in $\zeta,\ldots,\zeta^{p-1}$, so by the Fundamental Theorem on Symmetric Polynomials it is a symmetric polynomial with integer coefficients in the monic polynomial having $\zeta,\ldots,\zeta^{p-1}$ as its roots. But that polynomial is just $\Phi_p(x) = x^{p-1} + \cdots + 1$.

Now let $e(x)$ be the polynomial $e(x) = f(x)\cdots f(x^{p-1})$ and consider $S = \sum_{i=0}^{p-1} e(\zeta^i)$. On the one hand, we immediately see that

$$S = f(1)^{p-1} + (p-1)f(\zeta)f(\zeta^2)\cdots f(\zeta^{p-1}).$$

On the other hand, write $e(x) = \sum_n A_n x^n$. Since $f(x)$ has integer coefficients, $e(x)$ certainly has integer coefficients. Of course, $\zeta^n = 1$ whenever n is a multiple of p. Also, whenever n is not a multiple of p, ζ^n is a root of $\Phi_p(x)$ so $0 = \Phi_p(\zeta^n) = (\zeta^n)^{p-1} + \cdots + 1$. Hence we see that $S = \sum_{n \text{ a multiple of } p} A_n p$. Thus $f(1)^{p-1} + (p-1)f(\zeta)f(\zeta^2)\cdots f(\zeta^{p-1}) \equiv 0 \pmod{p}$ and the lemma follows.

Kronecker completes the proof as follows.

Suppose $\Phi_p(x)$ is not irreducible and write $\Phi_p(x) = f(x)g(x)$, a product of nonconstant polynomials. By Gauss's lemma, $f(x)$ and $g(x)$ both have integer coefficients. Then $p = \Phi_p(1) = f(1)g(1)$. One of these factors must be ± 1, so suppose $f(1) = \pm 1$. On the one hand, $f(\zeta^k) = 0$ for some k that is nonzero $\pmod{p}$ (as these are the roots of $\Phi_p(x)$), so $f(\zeta)\cdots f(\zeta^{p-1}) = 0$, but on the other hand $f(1)^{p-1} \equiv 1 \pmod{p}$, contradicting the above congruence.

The next step is due to Serret (Serret 1850), who proved the theorem in the prime power case. The strategy of his proof is similar to that of Kronecker's proof above, and he begins with a lemma that generalizes Kronecker's lemma.

Lemma: *Let p be a prime and let $f(x)$ be a polynomial with integer coefficients with $f(1) \equiv 1 \pmod{p}$. Then for any positive integer k,*

$$\prod f(\zeta) \equiv 1 \pmod{p}$$

where the product is taken over all primitive p^k-th roots of 1.

Proof: Let $F_n(x)$ be the polynomial $F_n(x) = f(x)f(x^2)\cdots f(x^{p^n})$. Let ζ_k be a (fixed but arbitrary) primitive p^k-th root of 1. Then $\zeta_{k-j} = \zeta_k^{p^j}$ is a primitive p^{k-j}-th root of 1 for $j = 0,\ldots,k$.

On the one hand, writing $F(x) = A_0 + A_1 x + \cdots$, we see just as before that

$$\sum F(\zeta) \equiv 0 \ (\text{mod } p^k),$$

where $\sum$ denotes the sum over all the p^k-th roots of 1.

On the other hand, since every p^k-th root of 1 is a primitive p^j-th root of 1 for exactly one value of j between 0 and k,

$$\sum F(\zeta) = \sum_k F(\zeta) + \sum_{k-1} F(\zeta) + \cdots + \sum_1 F(\zeta) + \sum_0 F(\zeta),$$

where $\sum_j$ denotes the sum over the primitive p^j-th roots of 1. Now for any $j = 1, \ldots, k$, there are exactly $p^j - p^{j-1}$ primitive p^j-th roots of 1, so

$$\sum F(\zeta) = (p^k - p^{k-1}) F_k(\zeta_k) + (p^{k-1} - p^{k-2}) F_{k-1}(\zeta_{k-1})^p + \cdots$$
$$+ (p^2 - p) F_2(\zeta_2)^{p^{k-2}} + (p - 1) F_1(\zeta_1)^{p^{k-1}} + f(1)^{p^k}.$$

Thus this sum is $\equiv 0 \ (\text{mod } p^k)$.

We now proceed by complete induction on k. For $k = 1$ this congruence is

$$(p - 1) F_1(\zeta_1) + f(1)^p \equiv 0 \ (\text{mod } p)$$

which immediately gives

$$F_1(\zeta_1) \equiv 1 \ (\text{mod } p)$$

and hence from the definition of the polynomial $F_1(x)$,

$$\prod f(\zeta) \equiv 1 \ (\text{mod } p)$$

in this case.

Now suppose $F_1(\zeta_1) \equiv 1 \ (\text{mod } p), \ldots, F_{k-1}(\zeta_{k-1}) \equiv 1 \ (\text{mod } p)$. Then the above congruence yields the congruence

$$-p^{k-1} F_k(\zeta_k) + (p^{k-1} - p^{k-2}) + (p^{k-2} - p^{k-3}) + \cdots + (p - 1) + 1 \equiv 0 \ (\text{mod } p^k).$$

(Here Serret uses without comment the following elementary fact, which is easy to prove by induction: If $a \equiv 1 \ (\text{mod } p)$ then $a^{p^j} \equiv 1 \ (\text{mod } p^{j+1})$ for every positive integer j.) This sum telescopes, yielding the congruence

$$-p^{k-1} F_k(\zeta_k) + p^{k-1} \equiv 0 \ (\text{mod } p^k)$$

which immediately gives

$$F_k(\zeta_k) \equiv 1 \pmod{p}.$$

Now

$$F_k(\zeta_k) = \prod f(\zeta) F_{k-1}(\zeta_{k-1})$$

so

$$\prod f(\zeta) \equiv 1 \pmod{p}$$

as claimed.

Given this lemma, the proof finishes as before. Suppose $\Phi_p(x)$ is not irreducible and write $\Phi_p(x) = f(x)g(x)$, a product of nonconstant polynomials. By Gauss's lemma, $f(x)$ and $g(x)$ both have integer coefficients. Then $p = \Phi_p(1) = f(1)g(1)$. One of these factors must be ± 1, so suppose, multiplying $f(x)$ and $g(x)$ by -1 if necessary, that $f(1) = 1$. Then $\prod f(\zeta) \equiv 1 \pmod{p}$, which is impossible as $f(x)$ has some primitive p^k-th root of 1 as a root, so this product is 0.

Kronecker proved the general case in 1854; we discuss it below. But he returned to the prime case in 1856 and gave a proof that is even simpler than his 1845 proof. The title of his paper (Kronecker 1856) only refers to the prime case, but the proof, as he observed in the last sentence of the paper, extends without essential change to the prime power case. For simplicity, we will follow him in just giving the proof in the prime case.

Suppose then that $\Phi_p(x)$ is not irreducible and write $\Phi_p(x) = f(x)g(x)$, a product of nonconstant polynomials. By Gauss's lemma, $f(x)$ and $g(x)$ both have integer coefficients. Then $p = \Phi_p(1) = f(1)g(1)$. One of these factors must be ± 1, so suppose $f(1) = \pm 1$.

For each $k = 1, \ldots, p-1$, let m_k be a positive integer with $km_k \equiv 1 \pmod{p}$. Let ζ be a root of $f(x)$. Then $f(\zeta^{km_k}) = f(\zeta) = 0$ so $x = \zeta^k$ is a root of the polynomial $f(x^{m_k})$, i.e., $x - \zeta^k$ divides $f(x^{m_k})$. Hence the product

$$e(x) = f(x^{m_1}) \cdots f(x^{m_{p-1}})$$

is divisible by

$$(x - \zeta) \cdots (x - \zeta^{p-1}) = \Phi_p(x) = x^{p-1} + \cdots + 1.$$

By Gauss's Lemma, the quotient $q(x) = e(x)/\Phi_p(x)$ is a polynomial with integer coefficients. In particular $q(1)$ is an integer. But $e(1) = f(1)^{p-1} = (\pm 1)^{p-1}$ while $\Phi_p(1) = p$, a contradiction.

(In case $n = p^r$ is a prime power, the argument goes through by letting k range over the integers from 1 to $n - 1$ that are relatively prime to p, and using the fact that $\Phi_{p^r}(x) = x^{p^{r-1}(p-1)} + x^{p^{r-1}(p-2)} + \cdots + x^{p^{r-1}} + 1$, from which once again we see $\Phi_{p^r}(1) = 1$.)

This is evidently both a simplification and a generalization of Kronecker's earlier argument, but it is in very much the same spirit. In particular, the polynomial we have denoted $e(x)$ is the same polynomial in each of those proofs.

(Note that this proof cannot be generalized further, as if n is not a prime power $\Phi_n(1) = 1$, as is easy to show and as Kronecker undoubtedly knew.)

One can only suppose that Kronecker's motivation for this 1856 proof was to give as simple a proof as possible. For the prime case, it is indeed simpler than his earlier proof, though not as simple as, and less general than, proofs given by Schönemann (in 1846) and Eisenstein (in 1850). For the prime power case, it is simpler than Serret's argument, and much simpler than his own proof for the general composite case.

As of 1846, only the prime case was known, by Gauss's original proof and Kronecker's simpler, but still complicated, proof. At that point Schönemann entered the story. He wrote a long paper (Schönemann 1846) in which he investigated polynomials with integer coefficients, and developed a simple and general criterion for such a polynomial $f(x)$ to be irreducible when reduced (mod p^2), where p is a prime. It is:

Lemma: *Let $f(x)$ be a polynomial of degree k with integer coefficients. Suppose that, for some prime p and some integer a,*

$$f(x) = (x - a)^k + pg(x)$$

for some polynomial $g(x)$ with integer coefficients with $g(a)$ not divisible by p. Then $f(x)$ is irreducible (mod p^2), (i.e., there do not exist polynomials with integer coefficients $h(x)$ and $k(x)$ with $f(x) \equiv h(x)k(x)$ (mod p^2)).

Schönemann explicitly states that he wants to show the power of his criterion (a clear case of benchmarking), and to do so he observes that his criterion applies to show that the (mod p^2) reduction of $\Phi_p(x)$ is irreducible: $\Phi_p(x) = (x - 1)^{p-1} + pg(x)$ for some polynomial $g(x)$ and, setting $x = 1$, $p = \Phi_p(1) = pg(1)$ so $g(1) = 1$. (He makes the first claim without further comment, but the justification for it is the observation that, by the binomial theorem, $(x - 1)^p = x^p + c_{p-1}x^{p-1} \cdots + \cdots + c_1 x + (-1)^p$ with all the coefficients $c_{p-1}, \ldots, c_1$ divisible by p. Hence $x^p - 1 = (x-1)^p + pg_0(x)$ for some polynomial $g_0(x)$. Setting $x = 1$ we see that $g_0(1) = 0$, so the polynomial $g_0(x)$ is divisible by the polynomial $x - 1$, say $g_0(x) = (x - 1)g(x)$ for some polynomial $g(x)$. Then

$$\Phi_p(x) = (x^p - 1)/(x - 1) = (x - 1)^{p-1} + pg(x).$$

as claimed.) This immediately implies (as he observes), that $\Phi_p(x)$ is irreducible over the integers, and hence, by Gauss's Lemma, over the rationals as well.

In 1850, Eisenstein appeared on the scene. In his paper (Eisenstein 1850) he investigated polynomials with coefficients in the "Gaussian integers" $\mathbb{Z}(i)$

rather than in the (ordinary) integers, and developed an irreducibility criterion. He too stated that he wanted to show the power of his criterion (another case of benchmarking), and so he applied it to the polynomial $\Phi_p(x)$ over the integers. His (well-known) criterion is:

Lemma: *Let $f(x) = a_n x^n + \cdots + a_1 x + a_0$ with integer coefficients. Suppose that, for some prime p, a_n is not divisible by p, $a_{n-1}, \ldots, a_1$ are all divisible by p, and a_0 is divisible by p but not by p^2. Then $f(x)$ is irreducible over the integers.*

Proof: Suppose that $f(x) = g(x)h(x)$ where $g(x)$ and $h(x)$ have integer coefficients. Let $g(x) = b_k x^k + \cdots + b_0$ and $h(x) = c_m x^m + \ldots + c_0$. Then neither b_k nor c_m is divisible by p. Since a_0 is divisible by p but not p^2, exactly one of b_0 and c_0 is divisible by p; suppose that one is b_0. We proceed inductively. First, $a_1 = b_0 c_1 + c_0 b_1$. Since a_1 and b_0 are divisible by p, and c_0 is not, b_1 is divisible by p. Next, $a_2 = b_0 c_2 + b_1 c_1 + b_2 c_0$. Since a_2, b_1, and b_0 are divisible by p, and c_0 is not, b_2 is divisible by p. Continuing in this fashion, we eventually conclude b_k is divisible by p, a contradiction.

The proof of this lemma could hardly be easier. It does not apply immediately to $\Phi_p(x)$, but, as Eisenstein observed, it can be made to do so by a simple trick. The polynomial $f(x)$ is irreducible if and only if the polynomial $f(x + 1)$ is. But

$$
\begin{aligned}
\Phi_p(x + 1) &= \frac{(x + 1)^p - 1}{(x + 1) - 1} \\
&= \frac{((x^p + px^{p-1} + \cdots + px + 1) - 1)}{x} \\
&= x^{p-1} + px^{p-2} + \cdots + p
\end{aligned}
$$

where, by the binomial theorem, all the omitted coefficients are divisible by p. Thus $\Phi_p(x + 1)$ is irreducible, and so $\Phi_p(x)$ is as well.

We have not given the proof of Schönemann's irreducibility criterion, as it is essentially the same as Eisenstein's. In fact, the two criteria are equivalent—either one may be easily derived from the other (and so they apply to exactly the same polynomials). As to why Eisenstein published a result that was essentially the same as Schönemann's, it seems that he was simply unaware of Schönemann's work and independently (re)discovered the result.

Now we turn to the general case. Just as the first proof in the prime case, that of Gauss, was quite complicated and was later simplified, the first proof in the general case, that of Kronecker (1854), was also quite complicated and was later simplified. We give that proof next.

Proof: Let $n = p_1^{k_1} p_2^{k_2} p_3^{k_3} \cdots$ be the prime factorization of n. Write $n = mp^k$ with m relatively prime to p. Kronecker proceeds inductively, first proving the following claim.

Theorem: *The polynomial* $\Phi_{p^k}(x) = x^{p^k} + x^{(p-1)p^{k-1}} + \cdots + x^p + 1$ *cannot be factored into two polynomials of lower degree whose coefficients are rational functions of the m-th roots of* 1, *with rational coefficients, as long as m is relatively prime to p.*

Proof: For simplicity, let $q = p^k$. Also, abbreviate polynomial with integer coefficients by integer polynomial.

Let $f(x) = \Phi_q(x)$ and let ζ be a primitive q-th root of 1. Suppose $f(x) = \varphi(x)\psi(x)$ where $\varphi(x)$ and $\psi(x)$ are polynomials with coefficients rational functions of α, where α is a primitive m-th root of 1 with p not dividing m. Since $f(x)$ is monic we may assume that $\varphi(x)$ and $\psi(x)$ are monic.

Write

$$\varphi(x) = (x - \zeta^{h_1})(x - \zeta^{h_2}) \cdots , \qquad \psi(x) = (x - \zeta^{i_1})(x - \zeta^{i_2}) \cdots .$$

Write

$$\varphi(1) = \frac{a_0 + a_1\alpha + \cdots + a_{r-1}\alpha^{r-1}}{M} = \frac{A(\alpha)}{M},$$

$$\psi(1) = \frac{b_0 + b_1\alpha + \cdots + b_{r-1}\alpha^{r-1}}{N} = \frac{B(\alpha)}{N},$$

where r is the degree of $\mu(x)$, the monic polynomial of lowest degree having α as a root (in fact, by induction, $\mu(x) = \Phi_m(x)$, but Kronecker does not use, or even mention, that fact), $a_0, \ldots, a_{r-1}$ and M are integers with M relatively prime to $a_0, \ldots, a_{r-1}$, and $b_0, \ldots, b_{r-1}$ and N are integers with N relatively prime to $b_0, \ldots, b_{r-1}$. Thus $A(x)$ and $B(x)$ are polynomials with integer coefficients.

(Here Kronecker uses without comment the fact that any rational function $p(\alpha)/q(\alpha)$ is a polynomial in α. To see this, note that, since $q(\alpha) \neq 0$, the polynomial $q(x)$ is relatively prime to the polynomial $\mu(x)$ (since $\mu(x)$, being of lowest degree, is irreducible), so there are polynomials $a(x)$ and $b(x)$ with $1 = q(x)a(x) + \mu(x)b(x)$. Setting $x = \alpha$, we obtain $1/q(\alpha) = a(\alpha)$ and then $p(\alpha)/q(\alpha) = p(\alpha)a(\alpha)$ is a polynomial.)

Since $f(1) = p$ we have

$$A(\alpha)B(\alpha) = pMN.$$

Consider the first factor $\varphi(x)$. Setting $x = 1$ in the definition of $\varphi(x)$ we have $\varphi(1) = (1 - \zeta^{h_1})(1 - \zeta^{h_2}) \cdots$. Then, by the binomial theorem, $\varphi(1)^q = pY(\zeta)$ for some integer polynomial $Y(x)$. (The only terms in $(1 - \zeta^h)^p$ that do not have a coefficient divisible by p are the first and the last, which give $1 + (-1)^q$, which is 0 for p odd and 2 for $p = 2$.)

Then the equation

$$(z - pY(\zeta))(z - pY(\zeta^2)) \cdots (z - pY(\zeta^q)) = 0$$

has $z = \varphi(1)^q$ as a root. Expand the left-hand side in powers of z. The coefficient of the highest power of z is 1. Every other term is symmetric in $\zeta, \ldots, \zeta^q$, that is, in the roots of $x^q - 1 = 0$, multiplied by powers of p. Thus (by the Fundamental Theorem on Symmetric Polynomials) the coefficients of that polynomial are integer polynomials in the coefficients of $x^n - 1$, hence integers themselves. So the equation takes the form

$$z^q + pu_1 z^{q-1} + p^2 u_2 z^{q-2} + \cdots + p^q u_q = 0$$

for some integers $u_1, \ldots, u_q$. Substitute in this equation the root $z = \varphi(1)^q = A(\alpha)^q / M^q$, solve for the high-order term and clear the denominator, to conclude that

$$A(\alpha)^{m^2} = pC(\alpha)$$

for some integer polynomial $C(x)$. Then, since $A(\alpha) = M\varphi(1)$, we see that there is an expression of this form for $A(\alpha)^s$ for any $s \geq m^2$. Choose j such that $p^j \geq m^2$ and $p^j \equiv 1 \pmod{m}$, which is certainly possible as p is relatively prime to m. We thus obtain an expression of the form

$$A(\alpha)^{p^j} = pD(\alpha)$$

for some integer polynomial $D(x)$. Referring to our expression for $A(\alpha)$, and using the multinomial theorem, we have

$$A(\alpha)^{p^j} = a_0^{p^j} + a_1^{p^j} \alpha^{p^j} + a_2^{p^j} \alpha^{2p^j} + \cdots + a_{r-1}^{p^j} \alpha^{(r-1)p^j} + pE(\alpha)$$

for some integer polynomial $E(x)$. But $p^j \equiv 1 \pmod{m}$, so $\alpha^{p^j} = \alpha$. Thus this expression becomes equal to

$$pD(\alpha) = A(\alpha)^{p^j} = a_0^{p^j} + a_1^{p^j} \alpha + a_2^{p^j} \alpha^2 + \cdots + a_{r-1}^{p^j} \alpha^{(r-1)} + pE(\alpha)$$

so

$$a_0^{p^j} + a_1^{p^j} \alpha + a_2^{p^j} \alpha^2 + \cdots + a_{r-1}^{p^j} \alpha^{(r-1)} = pE(\alpha) - pD(\alpha) = pG(\alpha)$$

for some integer polynomial $G(x)$. Now $G(x)$ is a polynomial of unknown degree. But α is a root of the polynomial $x^m - 1$, and is a root of the irreducible monic polynomial $\mu(x)$, so $\mu(x)$ divides $x^m - 1$, and therefore, by Gauss's Lemma, $\mu(x)$ is an integer polynomial. This implies that $G(\alpha)$ has a unique expression of the form

$$G(\alpha) = g_0 + g_1\alpha + \cdots + g_{r-1}\alpha^{r-1}.$$

($\mu(x)$ is monic, so by the division algorithm $G(x) = \mu(x)Q(x) + R(x)$ for unique integer polynomials $Q(x)$ and $R(x)$ with $R(x)$ a polynomial of degree less than r. Substitute $x = \alpha$ to obtain $G(\alpha) = R(\alpha)$. That shows existence; uniqueness follows, for if we had two distinct expressions, their difference would be a polynomial expression in α of degree less than r whose value is 0, contradicting the minimality of r.)

Comparing the two expressions we have derived, we see

$$a_0^{p^k} = pg_0, \; a_1^{p^k} = pg_1, \ldots, \; a_{r-1}^{p^k} = pg_{r-1}.$$

Hence each of the integers $a_0, \ldots, a_{r-1}$ is divisible by p, and so $A(x)/p$ is an integer polynomial.

Applying the same argument to the second factor $\psi(x)$, we see that $B(x)/p$ is also an integer polynomial. Hence so is their product $H(x) = (A(x)/p)(B(x)/p)$. From the equation $A(\alpha)B(\alpha) = pMN$ we then conclude

$$pH(\alpha) = MN.$$

As with $G(\alpha)$ above, $H(\alpha)$ has a unique expression in powers of α less than r,

$$H(\alpha) = h_0 + h_1\alpha + \cdots + h_{r-1}\alpha^{r-1}$$

with $h_0, \ldots, h_{r-1}$ integers. Then, substituting,

$$p(h_0 + h_1\alpha + \cdots + h_{r-1}\alpha^{r-1}) = MN.$$

By the uniqueness of the expression, we must have

$$ph_0 = MN, \quad h_1 = \cdots = h_{r-1} = 0.$$

In particular, at least one of M and N is divisible by p.

But, as we have seen, each of $a_0, \ldots, a_{r-1}$ and $b_0, \ldots, b_{r-1}$ is divisible by p, so either M is not relatively prime to $a_0, \ldots, a_{r-1}$ or N is not relatively prime to $b_0, \ldots, b_{r-1}$, a contradiction.[2]

---

[2]In general, if $\mathbb{F}$ is a field, then the degree $[\mathbb{F}(\gamma) : \mathbb{F}]$ of the extension $\mathbb{F}(\gamma)$ of $\mathbb{F}$ is equal to the degree of the (unique monic) polynomial of lowest degree with coefficients in $\mathbb{F}$ having γ as a root, or equivalently the (unique monic) irreducible polynomial over $\mathbb{F}$ having γ as a root. Thus if ζ_n and ζ_m denote primitive n-th and m-th roots of 1, respectively, this theorem shows that $[\mathbb{Q}(\zeta_n) : \mathbb{Q}(\zeta_m)] = p^{r-1}(p-1) = \varphi(p^r)$. Using induction and the multiplicativity of degrees of field extensions, this result immediately implies the irreducibility of $\Phi_n(x)$: $[\mathbb{Q}(\zeta_n) : \mathbb{Q}] = \varphi(n)$, and since $\Phi_n(\zeta_n) = 0$ and $\Phi_n(x)$ has degree $\varphi(n)$, then $\Phi_n(x)$ must be irreducible. However, Kronecker does not use this argument.

Given this claim, Kronecker proceeds to show that $\Phi_n(x)$ is irreducible, as follows:

Let the primitive n-th roots of 1 be $\theta_1, \ldots, \theta_t$, the primitive q-th roots of 1 be $\zeta_1, \ldots, \zeta_s$, and the primitive m-th roots of 1 be $\alpha_1, \ldots, \alpha_r$. Then for each k, $\theta_k = \zeta_i \alpha_j$ for some unique i, j (and $t = sr$). Thus

$$\Phi_n(x) = \prod_k (x - \theta_k) = \prod_{(i,j)} (x - \zeta_i \alpha_j) = \prod_j \left(\prod_i (x - \zeta_i \alpha_j) \right)$$

$$= f_1(x) \cdots f_r(x).$$

Let $f(x) = \prod_i (x - \zeta_i) = 1 + x^{p^{k-1}} + x^{2p^{k-1}} + \ldots + x^{(p-1)p^{k-1}}$. Note that the polynomial $f(x/\alpha_j)$ has the same roots as the polynomial $f_j(x)$ (namely $\zeta_1 \alpha_j, \ldots, \zeta_s \alpha_j$), so they differ by a constant factor. Thus

$$\Phi_n(x) \text{ is a constant multiple of } f(x/\alpha_1) \cdots f(x/\alpha_r).$$

We will show that $\Phi_n(x)$, of degree $t = sr$, is irreducible by showing that any nonconstant factor of it has degree t. Suppose that $\varphi(x)$ is a factor of $\Phi_n(x)$. As a factor of that product, it has a root in common with one of those polynomials. Let the root be $\zeta_1 \alpha_1$. Then $\varphi(\zeta_1 \alpha_1) = 0$ and $f(x/\alpha_1) = 0$. Since the polynomials $\varphi(x)$ and $f(x/\alpha_1)$ have a common root, those polynomials have a common factor $\varphi(\alpha_1, x)$, a nonconstant monic polynomial in x with coefficients rational functions of α_1. Write $f(x/\alpha_1) = \varphi(\alpha_1, x)\psi(\alpha_1, x)$ and substitute $x = \alpha_1 z$ to obtain

$$f(z) = \varphi(\alpha_1, \alpha_1 z)\psi(\alpha_1, \alpha_1 z).$$

But by the first step in the proof, $f(z)$ cannot have a factor of this form of lower degree. Hence $\varphi(\alpha_1, x) = f(x/\alpha_1)$ and so

$$\varphi(x) = f(x/\alpha_1)\psi(\alpha_1, x).$$

But now, because $\alpha_1, \ldots, \alpha_r$ are roots of the same polynomial $\mu(x)$, which, by the inductive hypothesis, is irreducible, we have that $\varphi(x) = f(x/\alpha_i)\psi(\alpha_i, x)$ for every $i = 1, \ldots, r$, i.e., that $\varphi(x)$ is divisible by $f(x/\alpha_1), \ldots, f(x/\alpha_r)$. (To see this, observe that the equality $\varphi(x) = f(x/\alpha_1)\psi(\alpha_1, x)$ is equivalent to a collection of identities of the form $\lambda(\alpha_1) = 0$, with $\lambda(x)$ a polynomial with rational coefficients, one for each coefficient of $\varphi(x)$. But $\lambda(\alpha_1) = 0$ is equivalent to the polynomial $\lambda(x)$ being divisible by the polynomial $\mu(x)$, which is equivalent, for each i, to $\lambda(\alpha_i) = 0$.) These polynomials have no common root (since their roots are roots of the polynomial $x^n - 1$, which has no multiple roots), and consequently

$$\varphi(x) \text{ is divisible by the product } f(x/\alpha_1) \cdots f(x/\alpha_r)$$

so $\varphi(x)$ has degree t, as claimed.

Finally, Kronecker remarks that, although this argument is for the inductive step, it also starts the induction, as we may choose $m = 1$ (whence $\alpha = 1$).

(Of course, the argument simplifies quite a bit in this case. Examining this simplification, we see it is an alternate to Kronecker's 1856 proof. Roughly speaking, the proofs have similar levels of difficulty—the key in each is to come up with the right polynomial to consider.)

Evidently because of the importance of this result, Kronecker chose to prove it separately. But in the same paper he showed how to generalize his argument further to prove the following result: *Let α be a root of a monic polynomial with integer coefficients whose determinant is relatively prime to n. Then $\Phi_n(x)$ is irreducible over the field $\mathbb{Q}(\alpha)$.* (Nowadays this quantity is called the discriminant. It is, by definition, the square of the product of the differences of the distinct roots of the polynomial. For the polynomial $x^m - 1$ it is, as Kronecker knew but did not bother to remark, equal to $\pm m^m$.)

Kronecker's proof was simplified a few years later by his Berlin colleague Arndt, who, in the introduction to his paper (Arndt 1859), entitled "Einfacher Beweis ..." (Simple proof ...) wrote that that theorem was so important that it deserved another proof.

Arndt began his proof with an observation similar to the one with which Gauss began his. To be precise, he strengthened Gauss's observation for the special case he needed.

Lemma: *Let $g(x)$ be an arbitrary polynomial with integer coefficients. Let $g(x)$ have roots $r_1, \ldots, r_m$. For a positive integer k, set $g_k(x) = (x - r_1^k) \cdots (x - r_m^k)$ (so that the roots of $g_k(x)$ are the k-th powers of the roots of $g(x)$). Suppose that $k = p^e$ is a power of the prime p. Then $g_k(x)$ has integer coefficients and furthermore $g_k(x) \equiv g(x) \pmod{p}$.*

Proof: Suppose that $k = p$. The coefficients of $g(x)$ are, up to sign, $s_i = s_i(r_1, \ldots, r_m)$, the elementary symmetric polynomials in the roots of $g(x)$. Similarly, the coefficients of $g_p(x)$ are, up to sign, $\tilde{s}_i = s_i(r_1^p, \ldots, r_m^p)$, the elementary symmetric polynomials in the roots of $g_p(x)$. Thus we must show

$$\tilde{s}_i \equiv s_i \pmod{p} \text{ for each } i.$$

By the multinomial theorem,

$$s_i^p = \tilde{s}_i + pq(r_1, \ldots, r_m)$$

for some polynomial $q(r_1, \ldots, r_m)$ with integer coefficients, hence some integer polynomial in $s_1, s_2, \ldots$. Hence $s_i^p \equiv \tilde{s}_i \pmod{p}$. But by Fermat's Little Theorem, $a^p \equiv a \pmod{p}$ for every integer a, and so

$$s_i \equiv s_i^p \equiv \tilde{s}_i \pmod{p}$$

as claimed.

Now proceed inductively. If $k = p^2$, then $g_{p^2}(x) \equiv g_p(x) \equiv g(x)$ (mod p), etc.

Arndt also proceeds by induction on the number of prime factors of n.

The first case is $n = p^e$ for some prime p. Suppose that $\Phi_n(x)$ is not irreducible. Then $\Phi_n(x) = f(x)g(x)$, a product of nonconstant monic polynomials of positive degrees a and b.

Consider the polynomial $f_n(x)$ (in the notation of the above lemma). On the one hand, every root ζ of $f(x)$ is an (primitive) n-th root of 1, so directly from its definition, $f_n(x) = (x - 1)^a$. On the other hand, by that lemma, $f_n(x) \equiv f(x)$ (mod p). Thus

$$f(x) = (x - 1)^a + pq(x)$$

for some polynomial $q(x)$ with integer coefficients. By the same logic,

$$g(x) = (x - 1)^b + pr(x)$$

for some polynomial $r(x)$ with integer coefficients. Multiplying, we obtain

$$\Phi_n(x) = (x - 1)^{a+b} + pq(x)(x - 1)^b + pr(x)(x - 1)^a + p^2 q(x)r(x).$$

Setting $x = 1$ and dividing the resulting equation by p, we obtain

$$1 = pq(1)r(1),$$

which is impossible.

Now for the inductive step. Let $n = p^e m$ with m relatively prime to p and assume that $\Phi_m(x)$ is irreducible. Set $h(x) = \Phi_n(x)$ and $k(x) = \Phi_m(x)$, for simplicity. Every root ζ of $h(x)$ is a primitive n-th root of 1, and so for any such ζ, ζ^{p^e} is a primitive m-th root of 1, i.e., a root of $k(x)$, and all occur equally often. Thus $h_{p^e}(x) = k(x)^c$ for some c. (Counting degrees, we see $c = p^e$.)

Suppose that $h(x)$ is not irreducible. Then $h(x) = f(x)g(x)$, a product of nonconstant monic polynomials of positive degrees. Then, by the same argument, $f_{p^e}(x)$ and $g_{p^e}(x)$ both divide $k(x)^c$. Since we are assuming that $k(x)$ is irreducible, we must have $f_{p^e}(x) = k(x)^a$ and $g_{p^e}(x) = k(x)^b$ for some positive integers a and b. Now, again by the above lemma, we have $f_{p^e}(x) \equiv f(x)$ (mod p) and similarly for $g_{p^e}(x)$. Thus

$$f(x) = k(x)^a + pq(x)$$

for some polynomial $q(x)$ with integer coefficients. By the same logic,

$$g(x) = k(x)^b + pr(x)$$

for some polynomial $r(x)$ with integer coefficients. Multiplying, we obtain

$$\Phi_n(x) = k(x)^{a+b} + pq(x)\Phi_m(x)^b + pr(x)\Phi_m(x)^a + p^2q(x)r(x).$$

Letting $x = \zeta$ be an arbitrary (but fixed) primitive m-th root of $\Phi_m(x)$, we obtain

$$z = \Phi_n(\zeta) = p^2q(\zeta)r(\zeta).$$

To determine the value of z, observe that not only is the polynomial $x^n - 1$ divisible by the polynomials $x^{n/p} - 1$ and $\Phi_n(x)$, but also by their product, since they have no roots in common. Thus

$$\frac{x^n - 1}{x^{n/p} - 1} = \Phi_n(x)s(x)$$

for some polynomial $s(x)$ with integer coefficients. Now $(x^n - 1)/(x^{n/p} - 1) = (x^{n/p})^{p-1} + (x^{n/p})^{p-2} + \cdots + x^{n/p} + 1$. Also, m divides n/p, so $\zeta^{n/p} = 1$. Thus, setting $x = \zeta$ in the above equation we obtain

$$p = zs(\zeta)$$

and then from the previous equation we obtain

$$1 = pq(\zeta)r(\zeta)s(\zeta).$$

Now, since $\Phi_m(x)$ is irreducible of degree $d = \Phi(m)$, the product $q(\zeta)r(\zeta)s(\zeta)$ can be expressed uniquely as $c_0 + c_1\zeta + \cdots + c_{d-1}\zeta^{d-1}$ for some rational numbers $c_0, \ldots, c_{d-1}$, and, because $\Phi_m(x)$ is monic, $c_0, \ldots, c_{d-1}$ are integers. But this yields

$$0 = -1 + pc_0 + pc_1\zeta + \cdots + pc_{d-1}\zeta^{d-1},$$

which (again by the uniqueness of the expression) yields $1 = pc_0$, which is impossible.

The proofs given so far are quite different in detail, but follow common threads. The next proof of the general case, due to Dedekind (1857), introduced a completely new approach.

Before giving Dedekind's proof, we remark that the proofs that follow all work directly over the field of rational numbers $\mathbb{Q}$ (or, more precisely, over the integers $\mathbb{Z}$). Except for Landau's, they proceed by showing that if $f(x)$ is any irreducible factor of $\Phi_n(x)$, ζ is any root of $f(x)$, and p is any prime not dividing n, then ζ^p is also a root of $f(x)$ (since any primitive n-th root of 1 may be obtained from any given one by successively taking prime powers). Landau shows that, in this situation, ζ^k is a root of $f(x)$ for any k relatively prime to n, but he does this in an indirect way, actually showing that for any such k, ζ^m is a root of $f(x)$ for some $m \equiv k \pmod{n}$.

Here then is Dedekind's proof.

Proof: Let $f(x)$ be an irreducible factor of $\Phi_n(x)$ of positive degree, and suppose that there is some ζ with $f(\zeta) = 0$, but $f(\zeta^p) \neq 0$ for some p not dividing n. By Gauss's Lemma, $f(x)$ has integer coefficients. Let $f_1(x) = f(x) = (x - \zeta_1)\cdots(x - \zeta_k)$, where $\zeta_1 = \zeta$, and consider $f_p(x) = (x - \zeta_1^p)\cdots(x - \zeta_k^p)$. Then, by the lemma above, $g(x) = f_p(x)$ is a polynomial with integer coefficients and $f_p(x) \equiv f_1(x)$ (mod p).

The polynomial $g(x)$ is also irreducible. To see this, suppose $g(x)$ has a proper factor $h(x)$ of positive degree. Let q be an integer with $pq \equiv 1$ (mod p), and observe that $\zeta^{pq} = \zeta$. Then $g_q(x) = f_{pq}(x)$ would be a polynomial of lower degree than $f(x)$ having ζ as a root, contradicting the irreducibility of $f(x)$.

Also, $g(x) \neq f(x)$ as $g(\zeta^p) = 0$ but $f(\zeta^p) \neq 0$.

Since $f(x)$ and $g(x)$ are distinct irreducible polynomials, each of which divides $m(x) = x^n - 1$, their product $f(x)g(x)$ divides $m(x)$.

Then $\overline{f}(x)\overline{g}(x)$ divides $\overline{m}(x)$, where the overline denotes the (mod p) reduction. But $\overline{g}(x) = \overline{f}(x)$ so $\overline{f}(x)^2$ divides $\overline{m}(x)$, which is impossible as $\overline{m}(x)$ and its formal derivative $\overline{m}'(x) = nx^{n-1}$ are relatively prime.

There is a simplification of this proof that uses exactly the same approach, but avoids the use of symmetric functions. That proof can be found in van der Waerden's famous (and influential) algebra text (van der Waerden 1931).

Proof: Let $f(x)$ be as above and let $g(x)$ be an irreducible polynomial with $g(\zeta^p) = 0$. Then ζ is a root of the polynomial $g(x^p)$. But $f(x)$ is irreducible, which implies that $f(x)$ divides $g(x^p)$. Reducing (mod p), we have that $\overline{f}(x)$ divides $\overline{g}(x^p)$. But by the multinomial theorem, $g(x^p) \equiv g(x)^p$ (mod p). Thus $\overline{f}(x)$ divides $\overline{g}(x)^p$, which implies that $\overline{f}(x)$ and $\overline{g}(x)$ have a common irreducible factor $\overline{h}(x)$. But then, by the same argument as above, $\overline{h}(x)^2$ divides $\overline{m}(x)$, which is impossible for the same reason.

A proof along different lines was given by Landau in 1929 (Landau 1929). He evidently wanted to give as simple a proof as possible, since his proof takes exactly eight lines in the original. However, Landau's style was famously terse, and to explicate his proof takes quite a bit longer. Here is Landau's proof.

Proof: While, strictly speaking, the degree of the 0 polynomial is undefined, we adopt the convention in this proof that the 0 polynomial has degree less than d for any positive integer d, to eliminate some cumbersome language.

Suppose that $f(x)$ is an irreducible polynomial with integer coefficients with $f(\zeta) = 0$, where ζ is some n-th root of 1. Let $f(x)$ have degree d. By the division algorithm for polynomials, for each $j = 0, \ldots, n-1$, $f(x^j) = f(x)q_j(x) + r_j(x)$ for unique polynomials $q_j(x)$ and $r_j(x)$, where $r_j(x)$ is a polynomial of degree less than d. Since the value of ζ^j only depends on j (mod n), we have a finite set of polynomials $\{r_0(x), \ldots, r_{n-1}(x)\}$ such that, for any integer j, $f(\zeta^j) = r(\zeta)$ for $r(x)$ some polynomial in this set. Also, if $s(x)$ is any polynomial of degree less than d with $s(\zeta^j) = 0$, then we must have $s(x) = r(x)$, as otherwise $s(x) - r(x)$ would be a nonzero polynomial of degree less than d which has ζ as a root, and that would contradict the irreducibility of $f(x)$.

In particular, for any prime p,

$$f(\zeta^p) = f(\zeta^p) - 0 = f(\zeta^p) - f(\zeta)^p = r(\zeta)$$

for some polynomial $r(x)$ in this set (more precisely, for $r(x) = r_j(x)$ where $j \equiv p \pmod{n}$, $0 \le j \le n - 1$). By the multinomial theorem, $f(x^p) \equiv f(x)^p \pmod{p}$, so $f(x^p) - f(x)^p = pg(x)$ for some polynomial $g(x)$ with integer coefficients. There is a unique polynomial $h(x)$ of degree less than d with $h(\zeta) = g(\zeta)$, the polynomial obtained from the division algorithm, writing $g(x) = f(x)u(x) + h(x)$ (where the uniqueness again comes from the irreducibility of $f(x)$). Thus $r(\zeta) = pg(\zeta) = ph(\zeta)$ so we must have $r(x) = ph(x)$, again because $r(x)$ and $ph(x)$ each have degree less than d. Thus all of the coefficients of $r(x)$ are divisible by p, so in particular, each coefficient is either 0 or has absolute value at least p. Hence if A is the largest absolute value of the coefficients of all of the polynomials $\{r_0(x), \ldots, r_{n-1}(x)\}$, we must have $r(x) = r_j(x) = 0$ (the 0 polynomial), for any prime $p > A$, so $f(\zeta^p) = r(\zeta) = 0$ for any such prime. Applying this argument repeatedly, we conclude that $f(\zeta^m) = 0$ for any integer m all of whose prime factors are greater than A.

Now let k be any integer relatively prime to n. Set

$$m = k + n \prod q \,,$$

where the product is taken over all primes $\le A$ that do not divide k. We claim that m is an integer of this form. To see this, let p be any prime with $p \le A$. There are two possibilities:

(1) If p divides k, then p is not one of the factors in the product, and p does not divide n (since k and n are relatively prime), so p divides the first term in the sum but not the second, and hence p does not divide m.
(2) If p does not divide k, then p is one of the factors in the product, so p divides the second term in the sum but not the first, and hence p does not divide m.

Now $m \equiv k \pmod{p}$, so

$$f(\zeta^k) = f(\zeta^m) = 0$$

for any integer k relatively prime to n; hence, $f(x) = \Phi_n(x)$ and $\Phi_n(x)$ is irreducible.

Schur (1929) countered Landau's proof with a one-page proof that was almost as simple, and which appeared on the next page of the same journal in which Landau's proof appeared. The idea of using the discriminant goes back to Kronecker, but Schur used that idea in a much more effective way (making more use of algebraic number theory).

Proof: Let $f(x)$ be an irreducible factor of $x^n - 1$ of positive degree, let ζ_1 be a root of $f(x)$, and let p be relatively prime to n. (Of course, $f(x)$ has integer coefficients, by Gauss's Lemma.) We claim that $\zeta = \zeta_1^p$ is also a root of $f(x)$. Suppose not. Now $f(z) = (x - \zeta_1) \cdots (x - \zeta_k)$ for some n-th roots $\zeta_1, \ldots \zeta_k$ of 1. Then $0 \neq f(\zeta) = (\zeta - \zeta_1) \cdots (\zeta - \zeta_k)$, so $f(\zeta)$ divides the discriminant $\pm n^n$ of the polynomial $x^n - 1$ (as the discriminant is the product of the squares of the differences of all of the roots of $x^n - 1$, and $f(\zeta)$ is the product of the differences of some of the roots). However, by the multinomial theorem, $f(x^p) \equiv f(x)^p \pmod{p}$, so $f(\zeta^p) \equiv f(\zeta)^p \equiv 0 \pmod{p}$, where this congruence is to be understood in the algebraic integers. But this congruence means that p divides $f(\zeta)$, so that would imply that p divides n^n (in the algebraic integers, and hence in the rational integers), which it does not.

(To make this exposition self-contained, we shall calculate the discriminant Δ of the polynomial $x^n - 1$, up to sign. Let $\zeta = \exp(2\pi i/n)$. The polynomial $h(x) = (x^n - 1)/(x - 1) = x^{n-1} + \ldots + 1$ has roots $\zeta, \ldots, \zeta^{n-1}$, so $h(x) = \prod_{k=1}^{n-1}(x - \zeta^k)$ and so $n = h(1) = \prod_{k=1}^{n-1}(1 - \zeta^k)$. Then

$$\Delta = \prod_{i<j}(\zeta^i - \zeta^j)^2 = \pm \prod_{i \neq j}(\zeta^i - \zeta^j) = \pm \prod_{i \neq j} \zeta^i (1 - \zeta^{j-i})$$

$$= \pm \prod_i \zeta^i \left(\prod_{k=1}^{n-1}(1 - \zeta^k) \right) = \pm \prod_i \zeta^i (n) = \pm n^n.$$

Schur simply assumes his readers know this fact.)

11.1 Summary

In this chapter we can see many of the rationales for giving alternative proofs that we have adduced. The theorem that the n-th cyclotomic polynomial is irreducible was first proved in the prime case, then the prime power case, then in general. Moreover, the latter is a very substantial generalization (not just the removal of some technical hypothesis), and an important one, in that the result for general n plays a foundational role in algebraic number theory and field theory. The first proofs for the prime and general cases (Gauss's and Kronecker's) were both quite involved, and so we see successors aiming to simplify those proofs and make them more perspicuous. In the prime case, we have the Schönemann-Eisenstein criterion, a criterion applicable to a large class of polynomials, not just to $\Phi(x)$ (for example, it immediately shows that the polynomial $x^n - p$ is irreducible for any n and any prime p), but we see Schönemann and Eisenstein each demonstrating the utility of the criterion by applying it to give an easy derivation of the irreducibility of $\Phi_p(x)$. Finally, in the later proofs of the irreducibility of $\Phi_n(x)$ in general, we see proofs from a variety of quite different standpoints.

References

Arndt, F.: Einfacher Beweis für die Irreductibilität einer Gleichung in der Kreistheilung. J. reine und angewandte Math. **56**, 178–181 (1859)

Dedekind, R.: Beweis für die Irreduktibilität der Kreisteilunggleichung. J. reine und angewandte Math. **54**, 27–30 (1857)

Eisenstein, F. G. M.: Über die Irreductibilität und einige andere Eigenschaften der Gleichung, von welcher die Theilung der ganzen Lemniscate abhängt. J. reine und angewandte Math. **39**, 160–179 (1850)

Gauss, C.-F.: *Disquisitiones Arithmeticae*. Leipzig (1801)

Kronecker, L.: Beweis dass für jede Primzahl p die Gleichung $1 + x + \cdots + x^{p-1} = 0$ irreductibel ist. J. reine und angewandte Math. **29**, 280 (1845)

Kronecker, L.: Mémoire sur les facteurs irréductibles de l'expression $x^n - 1$. J. math. pures et appliquées **19**, 177–192 (1854)

Kronecker, L.: Démonstration de l'irreductibilité de l'équation $x^{n-1} + x^{n-2} + \cdots + 1 = 0$, ou n désigne un nombre premier. J. math. pures et appliquées, series 2, **1**, 399–400 (1856)

Landau, E.: Über die Irreduzibilität der Kreisteilunggleichung. Math. Zeitschr. **29**, 462 (1929)

Schönemann, T.: Von denjenigen Moduln, welche Potenzen von Primzahlen sind. J. reine und angewandte Math. **32**, 93–105 (1846)

Schur, I.: Zur Irreduzibilität der Kreisteilunggleichung. Math. Zeitschr. **29**, 463 (1929)

Serret, J. A.: Sur une question de théorie des nombres. J. math. pures et appliquées **15**, 296 (1850)

van der Waerden, B. L.: *Moderne Algebra*. Springer Verlag, Berlin (1931)

Chapter 12
The Compactness of First-order Languages

12.1 Background

A **first-order formal language** $\mathcal{L}$ (with identity) is characterized by disjoint (possibly empty) sets $\mathbf{C}_{\mathcal{L}}$ of *constant symbols*, $\mathbf{R}^n_{\mathcal{L}}$ of n-place *relation symbols* ($n \geq 1$) and $\mathbf{F}^n_{\mathcal{L}}$ of n-place *function symbols* ($n \geq 1$) — collectively constituting the **non-logical** symbols of the language — together with the following logical symbols:

  (i) a denumerable set $V_{\mathcal{L}}$ of *variable symbols*
 (ii) the unary connective $\neg$ (denoting negation)
(iii) the binary connective $\vee$ (denoting disjunction)
 (iv) the binary connective $\wedge$ (denoting conjunction)
  (v) the existential quantifier $\exists$
 (vi) the universal quantifier $\forall$
(vii) the binary relation symbol $=$ (denoting identity).

The syntax of any such language is determined by the following stipulations: An *expression* is any finite concatenation of symbols of the language. Constant symbols and variable symbols are *terms*, as is any expression of the form $f\tau_1\tau_2\ldots\tau_n$, where f is an n-place function symbol and $\tau_1, \tau_2, \ldots, \tau_n$ are terms. *Atomic formulas* are expressions of the form $R\tau_1\tau_2\ldots\tau_n$ or $= \tau_1\tau_2$, where R is an n-place relation symbol and $\tau_1, \tau_2, \ldots, \tau_n$ are terms. An expression is a *formula* if and only if it is either

(a) an atomic formula
(b) of the form $\neg\phi$, where ϕ is a formula
(c) the form $\vee\phi_1\phi_2$, where ϕ_1 and ϕ_2 are formulas
(d) of the form $\wedge\phi_1\phi_2$, where ϕ_1 and ϕ_2 are formulas
(e) of the form $= \phi_1\phi_2$, where ϕ_1 and ϕ_2 are formulas
(f) of the form $\exists v\phi$, where v is a variable symbol and ϕ is a formula, or
(g) of the form $\forall v\phi$, where v is a variable symbol and ϕ is a formula.

© Springer International Publishing Switzerland 2015
J.W. Dawson, Jr., *Why Prove it Again?*, DOI 10.1007/978-3-319-17368-9_12

(Informally, formulas such as $\lor\phi_1\phi_2$ or $= \phi_1\phi_2$ are usually written as $\phi_1 \lor \phi_2$ or $\phi_1 = \phi_2$, parentheses are used to enhance readability of more complex formulas, $\phi \longrightarrow \psi$ is used to abbreviate the formula $\neg\phi \lor \psi$, and $\phi \longleftrightarrow \psi$ to abbreviate $(\phi \longrightarrow \psi) \land (\psi \longrightarrow \phi)$.)[1] Terms and formulas in a first-order language are given semantic interpretations in terms of *structures* for that language and *assignments* to its variable symbols. A structure $\mathfrak{A}$ for $\mathcal{L}$ consists of a non-empty set A, over whose elements the variables are understood to range,[2] as well as

(1) for each constant symbol c in $\mathbf{C}_\mathcal{L}$, a designated element c^A of A;
(2) for each relation symbol R in $\mathbf{R}_\mathcal{L}^n$, an n-ary relation R^A on A; and
(3) for each function symbol f in $\mathbf{F}_\mathcal{L}^n$, a function f^A from A^n into A.

An assignment to the variable symbols of $\mathcal{L}$ is a function α from $V_\mathcal{L}$ into A.

Given a structure $\mathfrak{A}$ for $\mathcal{L}$ and an assignment α, an element τ^A of A is then denoted by each term τ of $\mathcal{L}$. Specifically, if τ is a constant symbol c, then τ^A is c^A; if τ is a variable symbol v, τ^A is $\alpha(v)$; and if τ is $f\tau_1 \ldots \tau_n$, τ^A is defined recursively as $f^A(\tau_1{}^A, \ldots, \tau_n{}^A)$. In a similar fashion, formulas are given truth values relative to $\mathfrak{A}$ and α. Namely, $R\tau_1\tau_2 \ldots \tau_n$ is true (or *satisfied*) in $\mathfrak{A}$ under the assignment α if and only if $R^A(\tau_1{}^A, \tau_2{}^A, \ldots, \tau_n{}^A)$; $= \tau_1\tau_2$ is true in $\mathfrak{A}$ under the assignment α if and only if the elements $\tau_1{}^A$ and $\tau_2{}^A$ of A are identical; $\neg\phi$ is true in $\mathfrak{A}$ under the assignment α if and only if ϕ is not true in $\mathfrak{A}$ under the assignment α; $\lor\phi_1\phi_2$ is true in $\mathfrak{A}$ under the assignment α if and only if ϕ_1 is true in $\mathfrak{A}$ under the assignment α or ϕ_2 is true in $\mathfrak{A}$ under the assignment α or both are true in $\mathfrak{A}$ under the assignment α; $\land\phi_1\phi_2$ is true in $\mathfrak{A}$ under the assignment α if and only if both ϕ_1 and ϕ_2 are true in $\mathfrak{A}$ under the assignment α; $\exists v\phi$ is true in $\mathfrak{A}$ under the assignment α if and only if for some a in A, ϕ is true in $\mathfrak{A}$ under the assignment α', where $\alpha'(v) = a$ and $\alpha'(u) = \alpha(u)$ for each variable symbol u different from v; and $\forall v\phi$ is true in $\mathfrak{A}$ under the assignment α if and only if for every a in A, ϕ is true in $\mathfrak{A}$ under the assignment α', where $\alpha'(v) = a$ and $\alpha'(u) = \alpha(u)$ for each variable symbol u different from v.

A formula ϕ of $\mathcal{L}$ is a *sentence* if each variable symbol in it (if any) is bound by (lies within the scope of) a quantifier.[3] For such formulas, satisfaction in a structure $\mathfrak{A}$ does not depend upon a particular assignment: a sentence ϕ is satisfied in $\mathfrak{A}$ with the assignment α if and only if it is satisfied in $\mathfrak{A}$ with any assignment to the variable symbols (in which case $\mathfrak{A}$ is said to be a **model** for ϕ). A set S of sentences of $\mathcal{L}$ is then said to be **satisfiable** if there is a structure $\mathfrak{A}$ that is a model for each of the sentences in S (that is, one in which *all* of the sentences in S are satisfied). S is said

---

[1] In addition, in some treatments of first-order logic the symbols $\land$ and $\forall$ are not taken as basic logical symbols. Instead, $\phi_1 \land \phi_2$ is introduced as an abbreviation for $\neg(\neg\phi_1 \lor \neg\phi_2)$ and $\forall v\phi$ as an abbreviation for $\neg\exists v\neg\phi$.

[2] That the variables range over the *elements* of A rather than subsets of A is what is indicated by the adjective 'first-order.'

[3] For a formal definition of the distinction between free and bound variables, see any modern logic text.

to be **finitely satisfiable** if every finite subset of S is satisfiable, and a sentence ϕ is said to be a **logical consequence** of the set of sentences S if ϕ is satisfied in *every* structure that is a model of S. A sentence of $\mathcal{L}$ that is a logical consequence of the *empty* set of sentences (that is, one that is satisfied in every structure for $\mathcal{L}$) is said to be **logically valid**.

First-order languages arose as a means of formalizing mathematical deductions. Such a language becomes a **first-order logic** when certain of its logically valid sentences are designated as *logical axioms* and a finite, non-empty set of truth-preserving *rules of inference* is specified. Particular mathematical theories, such as Peano arithmetic, are then given first-order formalizations by designating certain other sentences as *non-logical axioms*. As usual, a proof in such a logic is a finite sequence of formulas, each of which is either an axiom or the result of applying one of the rules of inference to previous formulas in the sequence, and the last of which is the statement that is proved.

12.2 Completeness and compactness

In terms of the notions just defined, the **compactness theorem** states that for any set S of sentences in a first-order language $\mathcal{L}$, S is satisfiable if and only if S is finitely satisfiable. Equivalently, for any sentence ϕ and set of sentences S of $\mathcal{L}, \phi$ is a logical consequence of S if and only if it is a logical consequence of some finite subset of S. (Otherwise the set $S \cup \{\neg\phi\}$ of sentences of $\mathcal{L}$ would be finitely satisfiable, and hence satisfiable.)

For languages containing only denumerably many non-logical symbols, the compactness theorem was first proved by Gödel in his paper (Gödel 1930), a rewritten version of his doctoral dissertation (Gödel 1929) (in which the compactness theorem did *not* appear). The principal result in Gödel (1930) is the (denumerable) **completeness theorem**, which in its simplest form states that a sentence ϕ in a denumerable language $\mathcal{L}$ is logically valid if and only if it is provable from the axioms and rules of inference adopted in Whitehead and Russell's *Principia Mathematica* (Whitehead and Russell 1910).[4] Equivalently, every sentence of $\mathcal{L}$ is either satisfiable or is formally refutable (that is, its negation is provable) in Whitehead and Russell's system. In the latter form, Gödel first established the theorem for languages without the equality symbol and then extended it to languages including it. He then further extended the result to denumerable sets S of sentences, which he showed must either be satisfiable or must contain a finite subset whose conjunction is refutable.

---

[4]Gödel's proof is easily adapted to other well-known systems of logical axioms and rules of inference, such as that in Hilbert and Ackermann (1928), where the question of completeness for first-order logics was first posed.

Today the compactness theorem is usually deduced as an immediate consequence of the latter result (since any finite set of sentences whose conjunction is refutable must be unsatisfiable). That argument, however, involves recourse to the syntactic notion of refutability, whereas the compactness theorem refers only to the semantic notion of satisfiability. And in Gödel (1930) the argument is the other way around: completeness for a denumerable set of formulas $\{\psi_n\}$ is obtained as an immediate consequence of compactness.

Sketch of Gödel's proof: To prove the completeness theorem for a single sentence ϕ, Gödel began by recalling that every first-order formula is provably equivalent to one in prenex normal form (that is, one in which all quantifiers occur at the front, followed by a quantifier-free *matrix*), and then further restricted attention to prenex sentences K beginning with a block of universal quantifiers and ending with a block of existential quantifiers. He showed that if every such sentence is either refutable or satisfiable, so is every formula. He then defined the degree of K to be the number of paired $\forall\exists$ quantifier blocks in its prefix, and proved by induction that every formula K of degree 1 is either satisfiable or refutable and that if every K of degree k is either satisfiable or refutable, so is every K of degree $k + 1$.[5]

The proof of the base case $k = 1$ involves the construction of a sequence A_n of formulas that are conjunctions of quantifier-free formulas obtained from the matrix A of K by substituting sequences of new variables into it. To each A_n a corresponding *propositional* formula[6] B_n is then associated. Earlier, in his paper (Bernays 1926), Paul Bernays had shown that every such formula is either satisfiable by some assignment of truth values to its statement variables or is refutable on the basis of axioms for the propositional calculus (which form a subset of the logical axioms of first-order logic). Consequently, either some B_n is refutable in the propositional calculus or every B_n is satisfiable. In the former case Gödel showed that K is refutable from the logical axioms of *Principia Mathematica*; in the latter case, he constructed a model (with domain the set of all non-negative integers) that satisfied K.

It was in extending the completeness theorem to denumerable sets of formulas that Gödel stated and proved the denumerable compactness theorem (designated there simply as *Satz X*). Given a denumerable set S of first-order formulas, he noted that one can specify a set S' of prenex formulas of degree 1 (as defined above) such that the satisfiability of any subset of S' is equivalent to that of the corresponding subset of S. He then constructed, for each integer n, a first-order formula B_n that was a conjunction of the quantifier-free matrices $A_1, \ldots, A_{n-1}$ of the first $n - 1$

---

[5]In passing from degree k to degree $k + 1$ a new relation symbol is adjoined to the language, a device originally introduced by Thoralf Skolem in his paper (Skolem 1920).

[6]One built up by applying connectives to variables representing statements. Such propositional variables were included in the formulations of first-order logic (then called the 'restricted functional calculus') given in Whitehead and Russell (1910) and in Hilbert and Ackermann (1928). Since Gödel adopted the former as the underlying system in Gödel (1930), his proof there thus also established that the compactness property holds for denumerable sets of *propositional* formulas.

formulas of S', with new, distinct sequences of free variables substituted into each A_i. The existential closure of each such formula B_n[7] is a logical consequence of the first n formulas of S, so if every finite subset of S is satisfiable, so is every B_n; and therefore, by an argument similar to that used to prove that every formula of degree 1 is either satisfiable or refutable, Gödel concluded that if every B_n is satisfiable, so is every formula in S.

Two aspects of the proofs just sketched should be noted: they are not perspicuous, because the details of the constructions are rather intricate; and that for compactness depends on forming conjunctions of a finite number of formulas that constitute an initial segment of an infinite list, and so does not extend to a non-denumerable sequence of formulas (in languages containing a non-denumerable number of non-logical symbols).

12.3   Compactness for non-denumerable languages

Perhaps in part because the compactness theorem was employed in Gödel (1930) just as a lemma to prove completeness for denumerable sequences of first-order formulas, its significance was overlooked for nearly two decades by all workers in the field except A.I. Malcev, who, in his paper (Malcev 1936), first stated and proved that compactness held for *propositional* logics with non-denumerably many propositional variables.[8]

Malcev's proof of that fact was straightforward: Given a language $\mathcal{L}$ containing $\aleph_\alpha$ non-logical symbols, well-order the formulas of $\mathcal{L}$ (assuming the Axiom of Choice) in a sequence of minimal order-type ω_α. Then for each initial segment of that sequence there must, by the induction hypothesis, be some satisfying truth-valuation. Again invoking the Axiom of Choice, suppose that one such has been chosen for each $\lambda < \alpha$. Since each formula in the sequence contains only finitely many new propositional variables, at least one of the finitely many distinct assignments of truth values to those variables must be common to $\aleph_\alpha$ of the chosen (total) valuations. So at each stage $\lambda < \omega_\alpha$ one may choose a total truth-valuation that extends the partial truth-valuations already defined on all propositional variables encountered at earlier stages and that satisfies all formulas of order-index less than or equal to λ.[9]

Nowhere in Malcev (1936) did Malcev state the corresponding extension of the compactness theorem for first-order logic; he did, however, attempt to show that

---

[7]That is, the sentence $\exists v_{i_1} \ldots \exists v_{i_m} B_n$, where $v_{i_1}, \ldots, v_{i_m}$ are all the variables free in B_n, which is satisfied in a structure $\mathfrak{A}$ if and only if there is an assignment α under which B_n is satisfied in $\mathfrak{A}$.

[8]The history of the compactness theorem is a tangled one, described in detail for first-order logic in Dawson (1993).

[9]For other proofs of the compactness of propositional logic, see Paseau (2010).

for every set S of first-order formulas there was a corresponding set of formulas of the propositional calculus that were 'equivalent' to those of S, since it was 'well-known' that every first-order formula could be replaced by an 'equivalent' prenex formula of the form $\forall v_{i_1} \ldots \forall v_{i_n} \exists v_{j_1} \ldots \exists v_{j_m} \phi$. Then later, in his Russian-language paper (Malcev 1941), he *did* state the compactness theorem for non-denumerable sets of first-order formulas, for whose proof he referred readers to his 1936 paper.

The correctness of the argument in Malcev (1936) was subsequently disputed, turning on the question of just what 'equivalent' was supposed to mean. In the end, commentators concluded that the argument outlined there *could* be made to work, using a strengthened form of a theorem proved in Skolem (1920) (cf. footnote 6 above), if by 'equivalent' formulas Malcev meant that the satisfiability of either one followed from that of the other.[10]

A proof based on normal forms: Thirty years later, a proof of compactness for non-denumerable sets of first-order formulas, similar to that outlined in Malcev (1936) but involving a reduction to a different class of prenex formulas (a result also due to Skolem), was published in the textbook (Kreisel and Krivine 1971).

Specifically, in Skolem (1920) it was shown that to every formula ϕ in a first-order language $\mathcal{L}$ there corresponds a purely universal formula[11] $\hat{\phi}$ with the same free variables as ϕ, in a language $\mathcal{L}'$ obtained from $\mathcal{L}$ by adding new function symbols, such that every structure $\mathfrak{A}'$ for $\mathcal{L}'$ that satisfies $\hat{\phi}$ also satisfies ϕ, and every structure $\mathfrak{A}$ for $\mathcal{L}$ can be expanded to a structure $\mathfrak{A}'$ for $\mathcal{L}'$ in such a way that ϕ is satisfied in $\mathfrak{A}$ under an assignment α if and only if $\hat{\phi}$ is satisfied in $\mathfrak{A}'$ under that assignment. (The formula $\hat{\phi}$ is called the *Skolem normal form* of ϕ.) The proof of the compactness theorem for non-denumerable first-order languages given in Kreisel and Krivine (1971) then proceeds as follows:

(1) Compactness is first proved for non-denumerable sets of formulas in propositional logic, as above.

(2) It is shown that if a set S of purely universal first-order sentences has a model, then it has a model in which the domain is the set of all terms in the language whose function symbols are all those contained in the sentences of S, and in which the interpretation of those function symbols is the obvious one (that is, the value of the n-ary function symbol f at the n-tuple of terms $\tau_1, \ldots, \tau_n$ is the term $f\tau_1 \ldots \tau_n$.) Such a model is called a *canonical* model.

(3) Given terms $\tau_1, \ldots, \tau_n$ and a quantifier-free formula ϕ with free variables $v_1, \ldots, v_n$ in a first-order language $\mathcal{L}$, an easy argument shows that in any canonical model $\mathfrak{A}$, $\phi(\tau_1, \ldots, \tau_n)$ holds in $\mathfrak{A}$ if and only if $\phi(\tau_1, \ldots, \tau_n)$ is true in the propositional language whose statement letters are the atomic formulas of $\mathcal{L}$ when the atomic formulas in ϕ are given the valuation 'true.'

---

[10] See section 6 of Dawson (1993) for further details.

[11] That is, a prenex formula whose prefix contains no existential quantifiers.

Suppose then that S is a set of sentences in a first-order language $\mathcal{L}$, and that every finite subset $\{\phi_1, \ldots, \phi_n\}$ of S is satisfiable. Let $\hat{S}$ be the set of all sentences $\hat{\phi}$ in $\mathcal{L}'$, for ϕ in S. By Skolem's result, any model of $\{\phi_1, \ldots, \phi_n\}$ can be expanded to a model of $\{\hat{\phi}_1, \ldots, \hat{\phi}_n\}$, so $\hat{S}$ is finitely satisfiable; and since (again by Skolem's result) any model of $\{\hat{\phi}_1, \ldots, \hat{\phi}_n\}$ is a model of $\{\phi_1, \ldots, \phi_n\}$, it suffices to show that $\hat{S}$ is satisfiable.

Each sentence $\hat{\phi}$ in $\hat{S}$ is a purely universal sentence, with quantifier-free matrix A_ϕ, so consider $\mathcal{A} = \{A_\phi(\tau_1, \ldots, \tau_n) \mid \phi \text{ is in } \hat{S} \text{ and } \tau_1, \ldots, \tau_n \text{ are terms of } \mathcal{L}'\}$. $\mathcal{A}$ may be regarded as a set of propositional formulas in the propositional language whose statement letters are the atomic formulas of $\mathcal{L}'$. By (2), every finite subset of $\hat{S}$ has a canonical model. By (3), for every finite subset $\mathcal{B}$ of $\mathcal{A}$ there must therefore be a propositional truth-valuation that makes every sentence in $\mathcal{B}$ true. Hence, by the compactness theorem for propositional logic, there is a propositional truth-valuation T that makes every sentence in $\mathcal{A}$ true. Again invoking (3), any canonical structure corresponding to that truth-valuation[12] is then a model for $\hat{S}$.

The proof just given involves only semantic notions. It is more perspicuous than Gödel's and applies to languages of any cardinality; and as noted above, whereas in Gödel's proof new *relation* and *variable* symbols are adjoined to the language $\mathcal{L}$, in Kreisel and Krivine's proof new *function* symbols are adjoined instead.

A third approach, involving the adjunction of new *constant* symbols, was introduced by Leon Henkin in his doctoral dissertation (Henkin 1947). Also applicable to languages of any cardinality, it has become the standard technique for proving the completeness of first-order logic, from which compactness is then deduced as a corollary. But Henkin's method can also be used to prove compactness directly, without reference to completeness. (One text that does so is Hinman 2005). That argument, in outline, is as follows.

Henkin's proof: If $\mathcal{L}$ is a language whose set of non-logical symbols is of cardinality κ, then the set of all sentences of $\mathcal{L}$ is also of cardinality κ. Let S be any set of sentences of $\mathcal{L}$ that is finitely satisfiable. Then for any sentence ϕ of $\mathcal{L}$, either $S \cup \{\phi\}$ or $S \cup \{\neg\phi\}$ is finitely satisfiable. The goal is to show that S can be extended to a finitely satisfiable set of sentences H such that (i) for every sentence ϕ of $\mathcal{L}$, either ϕ is in H or $\neg\phi$ is in H, but not both, and (ii) for every sentence ϕ of $\mathcal{L}$ of the form $\exists v_i \chi$, if ϕ is in H, then there is a constant symbol c such that the sentence $\chi(c)$ is in H. (Such a set H is said to be *Henkin-complete*, and c is said to be a *Henkin-witness* for ϕ.) If so, then to establish compactness it suffices to show that any set of sentences that is finitely satisfiable and Henkin-complete is satisfiable.

To achieve that goal, a set of new constant symbols $\{d_\lambda \mid \lambda < \kappa\}$ is adjoined to $\mathcal{L}$ to form the extended language $\mathcal{L}'$. Using the Axiom of Choice, let $\{\phi_\lambda \mid \lambda < \kappa\}$

---

[12]That is, any canonical structure in which each relation symbol R that occurs in a formula in $\hat{S}$ is interpreted as the relation that holds of terms $\tau_1, \ldots, \tau_n$ if and only if T assigns the value 'true' to the atomic formula $R\tau_1\tau_2 \ldots \tau_n$.

be a fixed enumeration of the sentences of $\mathcal{L}$, and by transfinite recursion, for each $\lambda < \kappa$ define a set S_λ of formulas of $\mathcal{L}'$ as follows: Let $S_0 = S$, and let λ_0 be the least ordinal such that the constant d_{λ_0} does not occur in ϕ_λ nor in any formula in S_λ. If $S_\lambda \cup \{\phi_\lambda\}$ is satisfiable, let S_λ^0 be that set; otherwise, let it be S_λ. Then put $S_{\lambda+1}$ equal to $S_\lambda^0 \cup \{\chi(d_{\lambda_0})\}$ if ϕ_λ is in S_λ^0 and ϕ_λ is $\exists v_i \chi$, and equal to S_λ^0 otherwise. If λ is a limit ordinal, put $S_\lambda = \bigcup_{\sigma < \lambda} S_\sigma$. Finally, let $H = \bigcup_{\lambda < \kappa} S_\lambda$.

Clearly H contains S, and it is readily shown that H is finitely satisfiable and Henkin-complete. To construct a model for H, one defines an equivalence relation on the set of all constant symbols of $\mathcal{L}'$ by $c \sim d$ if and only if the sentence $c = d$ is in H, and takes the set of all equivalence classes under that relation to be the domain of a structure $\mathfrak{A}$ for $\mathcal{L}'$. If $\tilde{c}$ denotes the equivalence class of c, the interpretation of c in $\mathfrak{A}$ is just $\tilde{c}$; the interpretation in $\mathfrak{A}$ of an n-ary relation symbol R of $\mathcal{L}$ is the set of n-tuples $\widetilde{c_1}, \ldots, \widetilde{c_n}$ such that $Rc_1 \ldots c_n$ is in H; and the interpretation in $\mathfrak{A}$ of an n-ary function symbol f of $\mathcal{L}$ is that $f^{\mathfrak{A}}(\widetilde{c_1}, \ldots, \widetilde{c_n}) = \tilde{d}$ if and only if $fc_1 \ldots c_n = d$ is in H. (To show that these interpretations are well-defined, one checks that if $c_1 \sim c_1, \ldots, c_n \sim d_n$, then $Rc_1 \ldots c_n$ is in H if and only if $Rd_1 \ldots d_n$ is in H, and $fc_1 \ldots c_n = fd_1 \ldots d_n$ is in H.) One can then prove by induction on the formation of closed terms that for any closed term τ, $\tau^{\mathfrak{A}} = \tilde{d}$ if and only if $\tau = d$ is in H, and by induction on the formation of sentences that a sentence ϕ holds in $\mathfrak{A}$ if and only if ϕ is in H.[13] Hence $\mathfrak{A}$ is a model of H.

In the wake of Henkin's work, the idea of adding extra constant symbols to a language and invoking the compactness theorem became a basic technique for constructing models of first-order axiom systems. For example, add a single new constant symbol c to the language of Peano arithmetic and take S to be the set of sentences in that extended language obtained by adjoining to the first-order Peano axioms the denumerably many axioms $c > 0$, $c > s0$, $c > ss0$, $\ldots$ (where the function symbol s denotes the successor function). The compactness theorem then implies that S has a model, which must be a non-standard model of arithmetic, since the interpretation within it of the constant symbol c must be an element larger than every integer.

12.4   Algebraic proofs

All the proofs in the previous section make use of some form of the Axiom of Choice. The full strength of that axiom is not needed, but the compactness theorem for non-denumerable first-order languages is provably equivalent to the Boolean Prime Ideal Theorem, as well as to the Ultrafilter Theorem (stated and proved

---

[13]The completeness of propositional logic is invoked in showing that $\neg\phi$ holds in $\mathfrak{A}$ if and only if ϕ is not in H and that $\phi \vee \psi$ holds in $\mathfrak{A}$ if and only if ϕ is in H or ψ is in H.

below).[14] The latter is the basis for another proof of compactness that constructs a model for a finitely satisfiable set S of first-order sentences of any cardinality out of models for each of its finite subsets.

To carry out that proof, recall that a (proper) **filter** over a set I is a non-empty collection $\mathcal{F}$ of subsets of I which does not contain the null set and which is closed under the formation of supersets and finite intersections. (That is, if X is in $\mathcal{F}$ and $X \subset Y$, then Y is in $\mathcal{F}$, and if Y and Z are both in $\mathcal{F}$, then so is $Y \cap Z$.) If, in addition, for every subset X of I, either X or $I - X$ is in $\mathcal{F}$, then $\mathcal{F}$ is an **ultrafilter** over I.

From these definitions it follows immediately that the intersection of any finite number of elements of a filter is non-empty (that is, every filter has the *finite intersection property*). Conversely, any collection $\mathcal{C}$ of subsets of I that has the finite intersection property is contained within a filter over I, namely the filter $\mathcal{F} = \{Y \subseteq I \mid \text{for some } X_1, \ldots, X_n \text{ in } \mathcal{C}, \ X_1 \cap \cdots \cap X_n \subseteq Y\}$, called the filter *generated by* $\mathcal{C}$.

The **Ultrafilter Theorem** asserts that any such collection $\mathcal{C}$ is in fact contained within an ultrafilter over I. For with respect to the inclusion relation, the non-empty set of all filters over I that contain $\mathcal{C}$ satisfies the hypotheses of Zorn's Lemma, which asserts the existence of a maximal such filter, say $\mathcal{M}$. Then $\mathcal{M}$ must be an ultrafilter, since otherwise there would be a subset X of I such that neither X nor $I - X$ belonged to $\mathcal{M}$. But then $\mathcal{M} \cup \{X\}$ would satisfy the finite intersection property, and so be contained in some filter over I, contrary to the maximality of $\mathcal{M}$ (because if $Z \cap X$ were empty for some Z in $\mathcal{M}$, then Z would be a subset of $I - X$, so $I - X$ would be in $\mathcal{M}$).

Now suppose that S is a finitely satisfiable set of sentences in a first-order language $\mathcal{L}$. Let I be an index set for the collection F_S of finite subsets of S, and for each $i \in I$ choose a structure $\mathfrak{A}_i$, with domain A_i, in which the finite subset S_i of S is satisfied. For any ultrafilter $\mathcal{U}$ over I, let $\sim$ be the equivalence relation on $\prod_{i \in I} A_i$ defined by $F \sim G$ if and only if $\{i \mid F(i) = G(i)\} \in \mathcal{U}$. Let $\tilde{F}$ denote the equivalence class of F, and define an $\mathcal{L}$-structure $\mathfrak{A}_{\mathcal{U}}$ (the **ultraproduct** of the structures $\mathfrak{A}_i$ with respect to $\mathcal{U}$) as follows:

(a) The domain $A_{\mathcal{U}}$ of $\mathfrak{A}_{\mathcal{U}}$ is $\{\tilde{F} \mid F \in \prod_{i \in I} A_i\}$.
(b) The interpretation $c^{A_{\mathcal{U}}}$ of a constant symbol c of $\mathcal{L}$ is $\tilde{F}$, where $F(i) = c^{A_i}$ for each $i \in I$.
(c) The interpretation $R^{A_{\mathcal{U}}}$ of an n-ary relation symbol R of $\mathcal{L}$ holds of $\widetilde{F_1}, \ldots, \widetilde{F_n}$ if and only if $\{i \mid R^{A_i}(F_1(i), \ldots, F_n(i))\} \in \mathcal{U}$.
(d) For each n-ary function symbol f of $\mathcal{L}$, $f^{A_{\mathcal{U}}}(\widetilde{F_1}, \ldots, \widetilde{F_n}) = \tilde{G}$ if and only if $f^{A_i}(F_1(i), \ldots, F_n(i)) = G(i)$ for all $i \in I$.

Finally, if, for each $i \in I, \alpha_i$ is an assignment of elements of A_i to the variable symbols of $\mathcal{L}$, define the assignment α of elements of $A_{\mathcal{U}}$ to the variable symbols of $\mathcal{L}$ by $\alpha(v) = \widetilde{F_v}$, where $F_v(i) = \alpha_i(v)$.

---

[14]See Jech (1973), pp. 17–18, for the proofs of equivalence.

The compactness theorem is an easy consequence of the following fundamental theorem regarding ultraproducts.

Łoś's Theorem: For any set of $\mathcal{L}$-structures $\mathfrak{A}_i$ indexed by a set I, any family $\{\alpha_i\}$ of assignments to those structures, and any ultrafilter $\mathcal{U}$ on I, let the ultraproduct $\mathfrak{A}_{\mathcal{U}}$ and the associated assignment α be defined as above. Then:

(1) For every term τ of $\mathcal{L}$, the value $\tau^{\mathfrak{A}_{\mathcal{U}}}$ assigned to τ by α is $\tilde{F}$, where for every i in I, $F(i)$ is the value assigned to τ in $\mathfrak{A}_i$ under the assignment α_i.

(2) For every formula ϕ of $\mathcal{L}$, ϕ is satisfied in $\mathfrak{A}_{\mathcal{U}}$ under the assignment α if and only if $\{i \in I | \phi$ is satisfied in $\mathfrak{A}_i$ under the assignment $\alpha_i\} \in \mathcal{U}$. In particular, a sentence of $\mathcal{L}$ is satisfied in $\mathfrak{A}_{\mathcal{U}}$ if and only if $\{i \in I | \phi$ is satisfied in $\mathfrak{A}_i\} \in \mathcal{U}$.

The proof of (1) is by induction on the formation of terms, and that of (2) by induction on the length of formulas. For details, see, e.g., Hinman (2005), pp. 227–228.

To complete the proof of the compactness theorem, assume as above that for each i, the finite subset S_i of S is satisfied in $\mathfrak{A}_i$. For each $\mathcal{L}$-sentence ϕ, let $\mathcal{C}_\phi = \{S_i | i \in I$ and $\phi \in S_i\}$. Then the set $\mathcal{C}$ of all such $\mathcal{C}_\phi$ is a collection of subsets of F_S that satisfies the finite intersection property, because the intersection of $C_{\phi_1}, \dots, C_{\phi_n}$ must contain the set $\{\phi_1, \dots, \phi_n\}$. Hence, by the Ultrafilter Theorem, there is an ultrafilter $\mathcal{U}$ over F_S that contains $\mathcal{C}$. Now the set $M_\phi = \{S_i \mid \mathfrak{A}_i$ is a model of $\phi\}$ contains C_ϕ, since every ϕ in S_i is satisfied in $\mathfrak{A}_i$. Therefore each M_ϕ is in $\mathcal{U}$, so by (2) of Łoś's Theorem,[15] each sentence ϕ in S is satisfied in $\mathfrak{A}_{\mathcal{U}}$.

Note that unlike all the proofs of compactness considered earlier, the ultraproduct proof involves no expansion of the underlying language $\mathcal{L}$. Nor does it involve any reduction to the compactness of propositional logic, though that, too, can be proved using ultraproducts, by an argument similar to but simpler than that given above.

Like Henkin's method of constants, ultraproducts are a fundamental tool for constructing models. In particular, Abraham Robinson employed them to establish the consistency of non-standard analysis, thereby demonstrating that the infinitesimal methods of Leibniz could be given a rigorous foundation, and thus explain why those "ghosts of departed quantities" led to correct results. Beyond such latter-day justification of a historical practice, proponents of the non-standard approach also contend that arguments involving infinitesimals are more perspicuous to students of calculus than are $\epsilon - \delta$ computations. In the Introduction to Henle and Kleinberg (1979), for example, the authors declare "The power and beauty of [the non-standard approach] is sometimes astonishing. As Leibniz knew, the method of infinitesimals is the easy, natural way to attack ... [the] problems [of calculus], while the theory of limits represents the lengths to which mathematicians were willing to go to avoid them."

The role played by the finite intersection property in the ultraproduct proof of compactness recalls the definition of a compact space in topology (one in which

---

[15] With $i \in I$ replaced by $S_i \in F_S$, since $\mathcal{U}$ is an ultrafilter over F_S rather than I.

the intersection of a family of closed sets is empty if and only if the intersection of some finite subfamily is), and it was with that in mind that Tarski introduced the term 'compactness' as a name for the theorem that is the subject of this chapter. Specifically, for each structure $\mathfrak{A}$ for $\mathcal{L}$, let $Th(\mathfrak{A})$ (the *theory* of $\mathfrak{A}$) denote the set of all sentences that are true in $\mathfrak{A}$, and consider the collection S of all such theories. As noted in Keisler (1977), S can be made into a topological space (the *Stone space* of $\mathcal{L}$) by taking as basic closed sets those of the form $[\phi] = \{s \in S \,|\, \phi \in s\}$. A theory T is then satisfiable if and only if $\bigcap_{\phi \in T}[\phi]$ is non-empty, and the compactness theorem becomes the statement that the Stone space of $\mathcal{L}$ is a compact topological space.

In the denumerable case, the compactness theorem for first-order logic can also be established using ultrafilters on **Boolean algebras**, without resort to ultraproducts. Following Sikorski (1969), we take a Boolean algebra $\mathfrak{B}$ to be a structure of the form $< B, \oplus, \otimes, \bar{} >$, where B is a non-empty set, $\oplus$ and $\otimes$ (*join* and *meet*) are binary operations on the elements of B, and $\bar{}$ (*complement*) is a unary operation on the elements of B, such that join and meet are commutative and associative; each is distributive over the other; and for all a and b in B,

$$(a \otimes b) \oplus b = (a \oplus b) \otimes b = b,$$

(the so-called absorptive laws)   and

$$(a \otimes \bar{a}) \oplus b = (a \oplus \bar{a}) \otimes b = b.$$

It follows from those properties that for all a and b in B, $a \oplus a = a \otimes a = a$, $a \oplus \bar{a} = b \oplus \bar{b}$ and $a \otimes \bar{a} = b \otimes \bar{b}$. Writing $\mathbf{0}$ for $a \otimes \bar{a}$ and $\mathbf{1}$ for $a \oplus \bar{a}$, the last property above then becomes $\mathbf{0} \oplus b = \mathbf{1} \otimes b = b$. If $\mathbf{0} = \mathbf{1}$, then $B = \{\mathbf{0}\}$ and $\mathfrak{B}$ is said to be *degenerate*.

A partial ordering $\leq$ may be defined on B by stipulating that $a \leq b$ if and only if $a \oplus b = b$, or equivalently, by the absorptive laws, if and only if $a \otimes b = a$. With respect to $\leq$, $a \oplus b$ is then the supremum of a and b and $a \otimes b$ is the infimum of a and b; hence, by induction and the associative laws for $\oplus$ and $\otimes$, any finite set of elements $b_1, b_2, \ldots, b_n$ of B has supremum $b_1 \oplus b_2 \cdots \oplus b_n$ and infimum $b_1 \otimes b_2 \cdots \otimes b_n$.

In terms of the Boolean operations, a non-empty subset U of B is then an *ultrafilter on* B if and only if for all a and b in B: $\mathbf{0}$ is not an element of U; if a is an element of U and $a \leq b$, then b is an element of U; if a and b are elements of U, then so is $a \otimes b$; and $\bar{a}$ is an element of U if and only if a is not an element of U. Similarly, a subset A of B has the *finite intersection property* if and only if $a_1 \otimes \cdots \otimes a_n \neq \mathbf{0}$ for any finite number of elements $a_1, \ldots, a_n$ of A.[16] If A does have the finite intersection property, then the set

---

[16]That these definitions accord with those described above in the context of sets follows from the Stone Representation Theorem, which asserts that every Boolean algebra is isomorphic to a field of sets.

$\{b \in B | a_1 \otimes \cdots \otimes a_n \leq b$ for some $a_1, \ldots, a_n \in B\}$ is a filter on B that contains A. Any maximal such filter (under inclusion) is then an ultrafilter that contains A.

Given any set S of sentences of a first-order language $\mathcal{L}$, there are associated Boolean algebras whose elements are equivalence classes of formulas of $\mathcal{L}$. For the purpose of proving the compactness of $\mathcal{L}$, the appropriate equivalence relation is defined by $\phi \equiv \psi$ if and only if for some finite subset S_0 of S the formula $\phi \longleftrightarrow \psi$ is a logical consequence of S_0; that is, $\phi \longleftrightarrow \psi$ is satisfied under every assignment α in every structure $\mathfrak{A}$ for $\mathcal{L}$ that satisfies each sentence in S_0. Boolean operations are then well-defined by the equations $[\phi] \oplus [\psi] = [\phi \vee \psi], [\phi] \otimes [\psi] = [\phi \wedge \psi]$, and $\overline{[\phi]} = [\neg\phi]$, where $[\phi]$ denotes the equivalence class of the formula ϕ of $\mathcal{L}$. Let $\mathfrak{B}_S$ denote the resulting Boolean algebra.

If some finite subset of S is unsatisfiable (has no models), then $\phi \equiv \psi$ holds vacuously of all ϕ and ψ, so $\mathfrak{B}_S$ is degenerate. Otherwise S is finitely satisfiable. In that case, the equivalence classes $[\phi]$ and $[\neg\phi]$ must be distinct for any sentence ϕ, since no structure can satisfy both a sentence and its negation; so $\mathfrak{B}_S$ is non-degenerate.

To relate $[\exists x \phi]$ to $[\phi]$, it is necessary to consider suprema of denumerably infinite sets of elements of $\mathfrak{B}_S$. In general, such suprema may or may not exist; when the supremum of $b_1, b_2, \ldots$ does exist, it will be denoted by $\bigvee b_i$; the b_i themselves will be referred to as summands of such suprema. Similarly, if the infimum of $b_1, b_2, \ldots$ exists, it will be denoted by $\bigwedge b_i$. Then

Lemma: For any $a \in \mathfrak{B}_S$, $a \oplus \bigvee b_i = \bigvee (a \oplus b_i)$ and $a \otimes \bigvee b_i = \bigvee (a \otimes b_i)$.

Proof: Let $c = \bigvee b_i$. Then for all i, $b_i \leq c$, that is, $b_i \oplus c = c$. So

$$(a \oplus b_i) \oplus (a \oplus c) = (a \oplus a) \oplus (b_i \oplus c) = a \oplus c,$$

that is, $a \oplus b_i \leq a \oplus c$. Thus $a \oplus c$ is an upper bound for $\{a \oplus b_i\}$. Were it not the least upper bound, there would be a d such that for all $i, a \oplus b_i \leq d$, but $a \oplus c \not\leq d$, that is, $(a \oplus c) \oplus d \neq d$. But by definition of c, $c \leq d$, so $c \oplus d = d$, whence $a \oplus (c \oplus d) = a \oplus d$. Thus $a \oplus d \neq d$, that is $a \not\leq d$, contrary to $a \leq a \oplus b_i \leq d$.

To obtain the corresponding result for $\otimes$, note that the absorptive laws imply that $a \leq b$ is equivalent to $a \otimes b = a$ (for if $a \oplus b = b$ then $a = (a \oplus b) \otimes a = b \otimes a = a \otimes b$, and conversely if $a \otimes b = a$ then $b = (a \otimes b) \oplus b = a \oplus b$). So if $c = \bigvee b_i$, then $(a \otimes b_i) \otimes (a \otimes c) = (a \otimes a) \otimes (b_i \otimes c) = a \otimes b_i$ for all i, that is, $a \otimes b_i \leq a \otimes c$ for all i, whence $a \otimes c$ is an upper bound for $\{a \otimes b_i\}$.

To see that $a \otimes c$ is the least upper bound, note that $a \otimes b = a$ is also equivalent to $a \otimes \bar{b} = \mathbf{0}$. (For $a \otimes (b \oplus \bar{b}) = a \otimes \mathbf{1} = a = (a \otimes b) \oplus (a \otimes \bar{b})$, so if $a \otimes \bar{b} = \mathbf{0}$ then $a = a \otimes b$. Conversely, if $a = a \otimes b$, then $a \otimes \bar{b} = (a \otimes b) \otimes \bar{b} = a \otimes (b \otimes \bar{b}) = a \otimes \mathbf{0} = \mathbf{0}$.) Also, the distributive and commutative laws imply that for all a and b, $\overline{a \otimes b} = \bar{a} \oplus \bar{b}$ and $\overline{a \oplus b} = \bar{a} \otimes \bar{b}$. So if $a \otimes c$ were not the least upper bound, there would be some upper bound d for $\{a \otimes b_i\}$ such that $a \otimes c \not\leq d$. Hence $(a \otimes c) \otimes \bar{d} = c \otimes (a \otimes \bar{d}) \neq \mathbf{0}$. That is, $c \not\leq \overline{a \otimes \bar{d}}$, so $a \otimes \bar{d}$ is not an upper bound for $\{a \otimes b_i\}$. There must then be some b_j for which $b_j \not\leq \overline{a \otimes d}$. Consequently $b_j \otimes (a \otimes \bar{d}) = (a \otimes b_j) \otimes \bar{d} \neq \mathbf{0}$, so $a \otimes b_j \not\leq d$, contrary to the choice of d.

A basic notion in logic is that of changing quantified variables in a formula ϕ so as to obtain a logically equivalent formula (called a *variant* of ϕ) into which a given term τ may be substituted for one of the free variables without any of the finitely many variables in τ being 'captured' by one of the quantifiers. Since there are denumerably many variable symbols, all finite sets of variables may be ordered into a denumerable sequence. Then for each set $\{v_1, \ldots, v_k\}$ of variables such a formula ϕ' may be defined recursively on the structure of ϕ.[17]

In particular, for variables x and y, let $\phi'_x(y)$ denote the formula obtained by replacing all free occurrences of x by y in the first variant ϕ' of ϕ (ordered according to the ordering of finite sets of variables) in which y is freely substitutable for x. Then if S is finitely satisfiable, for any formula ϕ the value of $[\exists x \phi]$ in $\mathcal{B}_S$ is $\bigvee \{[\phi'_x(y)] | y \in V_{\mathcal{L}}\}$.

Proof: Note first that in $\mathcal{B}_S$, $[\phi] \leq [\psi]$ means $[\phi] \oplus [\psi] = [\phi \vee \psi] = [\psi]$, that is, $\phi \vee \psi \longleftrightarrow \psi$ is a logical consequence of some finite subset S_0 of S. But $\phi \longrightarrow \phi \vee \psi$ is logically valid for any formula ψ, so by *modus ponens*, $[\phi] \leq [\psi]$ holds if and only if $\phi \longrightarrow \psi$ is a logical consequence of S_0. Also, for any formula θ in which y is freely substitutable for x, $\theta_x(y) \longrightarrow \exists x \theta$ is logically valid. Hence $[\exists x \phi]$ is an upper bound for $\{[\phi'_x(y)] | y \in V_{\mathcal{L}}\}$. If $[\chi]$ is any other upper bound for $\{[\phi'_x(y)] | y \in V_{\mathcal{L}}\}$, then for each variable y, $\phi'_x(y) \longrightarrow \chi$ is a logical consequence of some finite subset S_y of S (the subscript indicating dependence on y). In particular, that holds for variables that do not occur in either ϕ or χ. If u is such a variable, then since u does not occur in ϕ, $\phi'_x(u)$ is just $\phi_x(u)$, the formula obtained from ϕ by replacing all free occurrences of x in ϕ by u. So $\phi_x(u) \longrightarrow \chi$ must hold in every structure $\mathfrak{A}$ that satisfies S_u (and since S is finitely satisfiable, such an $\mathfrak{A}$ must exist) under all assignments α. Now consider the formula $\exists x \phi \longrightarrow \chi$. If $\exists x \phi$ is not satisfied in $\mathfrak{A}$ under the assignment α, then $\exists x \phi \longrightarrow \chi$ is vacuously satisfied in $\mathfrak{A}$ under α. If $\exists x \phi$ *is* satisfied in $\mathfrak{A}$ under the assignment α, then ϕ is satisfied in $\mathfrak{A}$ under some assignment α' that differs from α at most in the value assigned to x (say a). So $\phi_x(u)$ is satisfied in $\mathfrak{A}$ under any assignment that differs from α' at most in assigning the value a to u, and hence also under any assignment β that differs from α at most in assigning the value a to u, because x does not occur free in $\phi_x(u)$. By *modus ponens*, χ is therefore satisfied in $\mathfrak{A}$ under β, and so also under α, since u does not occur in χ. Hence $\exists x \phi \longrightarrow \chi$ is satisfied under α. Consequently $[\exists x \phi] \leq [\chi]$.

A model for a finitely consistent set S of formulas can be constructed from any ultrafilter U on $\mathcal{B}_S$ that *preserves existential quantifiers*: one such that for all formulas ϕ and variables x, $[\exists x \phi] \in U$ if and only if for some y, $[\phi'_x(y)] \in U$;

---

[17] For details, see e.g. Hinman (2005), p. 111.

that is, $\bigvee\{[\phi_x'(y)]|y \in V_{\mathcal{L}}\} \in U$ if and only if $[\phi_x'(y)] \in U$ for some y. Since a denumerable language has only denumerably many formulas, the existence of such an ultrafilter is a consequence of the following result.

Theorem (Rasiowa-Sikorski): *For any Boolean algebra and any denumerably infinite set $\{s_k = \bigvee_i b_{ik}|k \in \omega\}$ of denumerably infinite sups, there is an ultrafilter U on $\mathfrak{B}$ that preserves all of them (that is, $s_k \in U$ if and only if some $b_{ik} \in U$).*

Proof: It suffices, for each $k \in \omega$, to find a summand $b_{i_k k}$ of s_k such that $\bigwedge_{j<k}(\overline{s_j} \oplus b_{i_j j}) \neq \mathbf{0}$. For then $\{\overline{s_k} \oplus b_{i_k k}|k \in \omega\}$ will have the finite intersection property and so be contained within an ultrafilter U. To see that any such U must preserve each of the suprema s_k, note first that $b_{i_k k} \leq s_k$ for every k. So $b_{i_k k} \in U$ implies $s_k \in U$. Conversely, suppose $s_k \in U$. Then as in the proof of the Lemma above, $b_{i_k k} \leq s_k$ is equivalent both to $s_k \otimes b_{i_k k} = b_{i_k k}$ and to $\overline{b_{i_k k}} \oplus s_k = \mathbf{1}$. Taking complements, the latter is also equivalent to $\overline{s_k} \otimes b_{i_k k} = \mathbf{0}$. Hence $b_{i_k k} = \mathbf{0} \oplus b_{i_k k} = \mathbf{0} \oplus (s_k \otimes b_{i_k k}) = (\mathbf{0} \oplus s_k) \otimes (\mathbf{0} \oplus b_{i_k k}) = s_k \otimes [(\overline{s_k} \oplus b_{i_k k}) \oplus b_{i_k k}] = s_k \otimes [\overline{s_k} \oplus b_{i_k k}] \in U$.

The $b_{i_k k}$ are defined recursively. Having defined $b_{i_j j}$ for all $j < k$ so that $c_k = \bigwedge_{j<k}(\overline{s_j} \oplus b_{i_j j}) \neq \mathbf{0}$, suppose that $c_k \otimes (\overline{s_k} \oplus b_{ik}) = \mathbf{0}$ for every summand b_{ik} of s_k. Then by the Lemma,

$$c_k = c_k \otimes \mathbf{1} = c_k \otimes (\overline{s_k} \oplus \bigvee_i b_{ik}) = c_k \otimes \bigvee_i (\overline{s_k} \oplus b_{ik}) = \bigvee_i c_k \otimes (\overline{s_k} \oplus b_{ik}) = \mathbf{0},$$

which contradicts the induction assumption. Therefore a $b_{i_k k}$ as desired must exist.

To construct a model $\mathfrak{M}$ for a finitely consistent set S of sentences of $\mathcal{L}$, let U be an ultrafilter on $\mathfrak{B}_S$ that preserves existential quantifiers. Define an equivalence relation $\simeq$ on the variables of $\mathcal{L}$ by the condition $x \simeq y$ if and only if $[x = y] \in U$. Let $< x >$ denote the equivalence class of x under $\simeq$, and take the domain of $\mathfrak{M}$ to be the set of all such equivalence classes. The constant, relation and function symbols are then interpreted in the obvious ways:

$$c^{\mathfrak{M}} = < x > \text{ if and only if } [c = x] \in U$$

$$R^{\mathfrak{M}}(< x_1 >, \dots, < x_n >) \text{ if and only if } [Rx_1 \dots x_n] \in U$$

$$f^{\mathfrak{M}}(< x_1 >, \dots, < x_n >) = < y > \text{ if and only if } [fx_1 \dots x_n = y] \in U.$$

In addition, let α be the assignment defined by $\alpha(x) = < x >$. Then U coincides with the filter $\{[\phi]|\phi \text{ is satisfied in } \mathfrak{M} \text{ under the assignment } \alpha\}$.

The proof that ϕ is satisfied in $\mathcal{M}$ under the assignment α if and only if $[\phi] \in U$ is by induction on the length of formulas. The fact that U preserves existential quantifiers is needed to show that $\exists x\phi$ is satisfied in $\mathcal{M}$ under α if and only if $[\exists x\phi] \in U$. For $\exists x\phi$ is satisfied in $\mathcal{M}$ under α if and only if for some y, ϕ is satisfied in $\mathcal{M}$ by an assignment that differs from α at most in assigning $< y >$ to x, which holds if and only if $\phi_x'(y)$ is satisfied in $\mathcal{M}$; and since the length of $\phi_x'(y)$ is less than that of $\exists x\phi$, the latter holds if and only if $[\phi_x'(y)] \in U$.

Note that since a first-order language has only denumerably many variable symbols, the domain of $\mathcal{M}$ is denumerable. Consequently, the Boolean proof cannot be extended to non-denumerable languages, in which a finitely consistent set S of sentences may have no denumerable models.

Like the method of constants and that of ultraproducts, Boolean algebras also have wider applications in logic. In particular, Boolean methods can be used to construct models of set theory that provide alternative proofs of independence results first derived by Paul Cohen's method of forcing.

12.5 Summary

This chapter has examined five proofs of the compactness theorem, which differ methodologically in fundamental respects. All are direct; that is, they do not derive compactness as a corollary of completeness. Two apply only to denumerable languages; the others apply to languages of any cardinality. In constructing a model for S, the algebraic proofs do not alter the language; the others do, in different ways. The ultraproduct proof also makes no reference to syntactic notions, but constructs a model for S directly from models for each of its finite subsets; and since the statement of the compactness theorem refers only to the semantic notion of satisfiability, the ultraproduct proof thus exhibits methodological purity.

References

Bernays, P.: Axiomatische Untersuchung des Aussagen-Kalkuls der *Principia Mathematica*. Math. Zeitsch. **25**, pp. 305–320 (1926)

Dawson, J.: The compactness of first-order logic: from Gödel to Lindström. Hist. and Philos. Logic **14**, pp. 15–37 (1993)

Feferman, S. et al. (eds): Kurt Gödel: Collected Works, vol. I. Oxford U.P., Oxford (1986)

Gödel, K.: Über die Vollständigkeit des Logikkalkuls. Doctoral dissertation, U. Vienna (1929). In (Feferman et al. 1986), pp. 61–101.

Gödel, K.: Die Vollständigkeit der Axiome des logischen Funktionenkalküls. Monatsh. Math. und Physik **37**, pp. 349–360 (1930). In (Feferman et al. 1986), pp. 103–123.

Henkin, L.: The completeness of formal systems. Doctoral dissertation, Princeton University (1947)

Henle, J., Kleinberg, E.: Infinitesimal Calculus. MIT Press, Cambridge, Mass. (1979)

Hilbert, D., Ackermann, W.: Grundzüge der theoretischen Logik. Springer, Berlin (1928)

Hinman, P.: Fundamentals of Mathematical Logic. A K Peters, Wellesley, MA (2005)

Jech, T.: The Axiom of Choice. North-Holland, Amsterdam and London (1973)

Keisler, H.J.: Fundamentals of model theory. In J. Barwise, Handbook of Mathematical Logic, pp. 47–103. North-Holland, Amsterdam New York and London (1977)

Kreisel, G., Krivine, J-L.: Elements of Mathematical Logic (Model Theory). North-Holland, Amsterdam (1971)

Malcev, A.I.: Untersuchungen aus dem Gebiete der mathematischen Logik. Math. Sbornik n.s. **1**, pp. 323–336 (1936)

Malcev, A.I.: [On the faithful representation of infinite groups by matrices] (in Russian). Math. Sbornik n.s. **8(50)**, pp. 405–422 (1941)

Paseau, A. Proofs of the compactness theorem. Hist. and Philos. Logic **31**, pp. 73–98 (2010). Corrigendum, Hist. and Philos. Logic **32**, p. 407 (2011)

Sikorski, R.: Boolean Algebras, 3rd ed. Springer, New York (1969)

Skolem, T.: Logisch-kombinatorische Untersuchungen über die Erfüllbarkeit oder Beweisbarkeit mathematischer Sätze nebst einem Theoreme über dichte Mengen. Skrifter utgit av Videnskaps-selskapet i Kristiana, I. Matematisk-naturvidenskabelig klasse **4**, pp. 1–36 (1920)

Whitehead, A.N., Russell, B.: Principia Mathematica, vol 1. Cambridge U.P., Cambridge (1910)

Chapter 13
Other Case Studies

I hope that the case studies in the preceding chapters have convinced the reader that the comparative study of alternative proofs is a worthwhile endeavor and that the informal criteria for distinguishing proofs described in Chapter 1 serve that purpose well. I hope too that some of the proofs discussed in those chapters will have been new to most readers, who will have found them to possess both intrinsic interest and pedagogical value.

This final chapter highlights some other theorems whose proofs have been the subject of comparative studies by other scholars. In addition, some theorems whose alternative proofs appear to be worthy subjects for further investigation are indicated.

13.1 Heron's formula

Heron's formula expresses the area of a triangle in terms of the lengths a, b, c of its three sides. Letting s denote the semiperimeter $\frac{1}{2}(a+b+c)$ and A the area, it states that $A^2 = s(s-a)(s-b)(s-c)$. In his books Dunham (1990, 1999) and articles Dunham (1985, 2011), William Dunham has presented five proofs of that formula, due to Heron of Alexandria, Isaac Newton, Leonhard Euler, Dunham himself, and Bernard Oliver. Here we do not give full details of all those proofs, but summarize their basic structure and highlight the similarities and differences among them and two proofs often found in textbooks, based, respectively, on the Law of Cosines and the Pythagorean Theorem.

Of the seven proofs, Heron's and Euler's are geometric; Newton's and both textbook proofs are primarily algebraic; while Dunham's and Oliver's are primarily trigonometric. Four of the proofs — those by Heron, Euler, Dunham, and Oliver — start with the same geometric construction: finding the *incenter* of the given triangle ABC (drawn so that its base is the longest side of the triangle). Specifically, since

© Springer International Publishing Switzerland 2015
J.W. Dawson, Jr., *Why Prove it Again?*, DOI 10.1007/978-3-319-17368-9_13

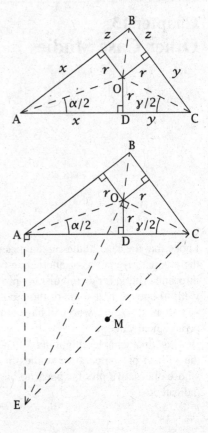

Fig. 13.1 Initial diagram for Heron's, Euler's, Dunham's, and Oliver's proofs

Fig. 13.2 Final diagram for Heron's proof

the bisector of any angle is the locus of points equidistant from the lines forming that angle, the intersection of any two bisectors of the angles of a triangle must be equidistant from all three sides of the triangle, and so must lie on the third angle bisector as well. The point O of intersection of those bisectors is thus the center of a circle inscribed within the triangle. Denoting the radius of that circle by r, the triangle is then seen to be composed of three pairs of congruent right triangles, each having altitude r and base (x, y or z) that is half of one of the sides. (See Figure 13.1.) Then $a = s - x, b = s - z, c = s - y$ and the area of the triangle is $r(x + y + z) = rs$.

From that point on those four proofs diverge. Heron adds three auxiliary line segments to the triangle: $\overline{AE}$ perpendicular to $\overline{AC}$, $\overline{OE}$ perpendicular to $\overline{OC}$, and $\overline{CE}$ (with midpoint M) (Figure 13.2). Euler extends the segments $\overline{CO}$ and $\overline{DO}$ and then constructs the segment through A perpendicular to the extension of $\overline{CO}$, meeting it at E and meeting the extension of $\overline{DO}$ at F (Figure 13.3). Dunham merely adds the altitude from vertex B to the triangle, and Oliver adds nothing at all.

The purpose of the auxiliary lines in Heron's diagram is not at all apparent at the outset. In fact, Heron used them to show that the quadrilateral $AOCE$ is inscribed within a semicircle centered at M, and then concluded from Proposition III,22 of

Fig. 13.3 Final diagram for
Euler's proof

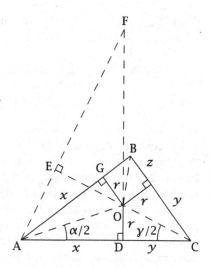

the *Elements* that angles *AEC* and *AOC* are supplementary. From that, in turn, he deduced that three pairs of triangles in the diagram were similar, and used ratios of their corresponding sides finally to conclude that $r^2s = xyz = (s-a)(s-c)(s-b)$, from which the formula bearing his name follows. His proof is ingenious, but as Oliver says, it is circuitous and "exhibits none of the symmetry of the problem or of the result," which it serves to *verify* rather than *explain* (Oliver 1993, p. 162).

Euler's argument is much more efficient and uses only basic facts about triangles. For in Figure 13.3, angle *AOE* is exterior to triangle *AOC* and complementary to angle *EAO*, so $\angle AOE = \alpha/2 + \gamma/2 = \pi/2 - \beta/2$ and $\angle EAO = \beta/2$. Consequently *OAE* and *OBG* are similar right triangles, as are *FOE*, *FAD*, *CAE*, and *COD*. Combining ratios of their corresponding sides then leads to $xyz = r^2s$, as in Heron's proof. Like that proof, Euler's commands assent that the result is true. But it still leaves the question: "What led him to draw those particular auxiliary lines?"

In Dunham's proof, letting h denote the length of the altitude dropped from B, we have immediately

$$\sin\alpha = \frac{h}{x+z}, \quad \sin(\alpha/2) = \frac{r}{\sqrt{x^2+r^2}} \quad \text{and} \quad \cos(\alpha/2) = \frac{x}{\sqrt{x^2+r^2}}.$$

The double-angle identity for sine then gives $\dfrac{h}{x+z} = \dfrac{2rx}{x^2+r^2}$. Solving for h and substituting the result obtained into the usual area formula yields the expression $(x+y)(x+z)\dfrac{rx}{r^2+x^2}$ for the area. Setting that equal to rs, the equation reduces once again, after some algebraic simplification and replacement of $x+y+z$ by s, to $r^2s = xyz$.

Oliver's argument is even simpler. Since $\alpha + \beta + \gamma = \pi$, the angles $\alpha/2$ and $\beta/2 + \gamma/2$ are complementary. Hence $\tan(\alpha/2) = \cot(\beta/2 + \gamma/2)$, so

$\tan(\alpha/2)\tan(\beta/2 + \gamma/2) = 1$. The summation identity for tangent then yields the symmetric equation

$$\tan(\alpha/2)\tan(\beta/2) + \tan(\alpha/2)\tan(\gamma/2) + \tan(\beta/2)\tan(\gamma/2) = 1.$$

That is, $\dfrac{r}{x}\dfrac{r}{z} + \dfrac{r}{x}\dfrac{r}{y} + \dfrac{r}{z}\dfrac{r}{y} = 1$. Multiplying both sides of that equation by xyz, $r^2(y + z + x) = r^2 s = xyz$.

For economy of means, Oliver's proof is unmatched. It uses no symbols except those in the triangle with angle bisectors, and the final algebra step is trivial. In contrast, the computations in the three algebraic proofs are much messier, involving several factorizations of a difference of two squares.

Those three proofs are in fact only minor variants of one another. Newton's, published in his *Arithmetica Universalis*, began with the diagram in Figure 13.4, in which the altitude h and a semicircle centered at A with radius c was added to the triangle ABC, whose base was extended to meet that semicircle externally at F. He then applied the Pythagorean Theorem to triangles ADB and CDB to obtain two different expressions for h^2. Equating them, and letting M denote the midpoint of the base b, he found that $|\overline{MD}| = (c^2 - a^2)/2b$ and $|\overline{DE}| = |\overline{ME}| - |\overline{MD}| = [2bc - (b^2 + c^2 - a^2)]/2b$. He then noted that if F were joined to B and B to E, h would also be the altitude of the *right* triangle FBE, whence FBE is similar to BDE. Equating ratios of their corresponding sides then gave

$$h^2 = |\overline{DE}||\overline{FD}| = \frac{[2bc - (b^2 + c^2 - a^2)][2bc + (b^2 + c^2 - a^2)]}{4b^2}.$$

He completed the proof by substituting the resulting value for h into the area formula $bh/2$, obtaining $[4b^2c^2 - (b^2 + c^2 - a^2)]^{1/2}/4$ for the area — which, after much tedious algebra, reduces to Heron's formula.

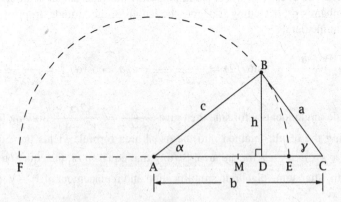

Fig. 13.4 Diagram for Newton's proof

But to obtain the equation displayed above, there is no need to introduce any lines to the triangle except the altitude. Indeed, the first of the common textbook proofs obtains two expressions for h^2 by applying the Pythagorean Theorem to each of the triangles ABD and BCD. Equating them yields an expression for $|\overline{AD}|$ that, when substituted back into one of the two Pythagorean equations, gives the same equation.

The second of the common textbook proofs instead employs the Law of Cosines, which gives $\cos \alpha = (b^2 + c^2 - a^2)/2bc$. Since $h = c \sin \alpha$, solving for $\sin \alpha$ and substituting into the area formula then again yields that the area is $[4b^2c^2 - (b^2 + c^2 - a^2)]^{1/2}/4$.

One is left to wonder why few of today's textbooks give the simpler proofs of Dunham or Oliver. Perhaps it is merely because in most U.S. high-school curricula algebra and geometry are taught before trigonometry.

13.2 Euler's formula for polyhedra

As recounted in Cromwell (1997), it was in a letter to Christian Goldbach dated November 1850 that Leonhard Euler announced his discovery of the formula $V - E + F = 2$ relating the number of vertices, edges, and faces of certain families of solids bounded by plane faces. He hoped to prove that the result held for arbitrary polyhedra, and in a paper published the next year in the proceedings of the St. Petersburg Academy he outlined a putative proof that it did. His idea was to lop off pyramidal corners of a polyhedron one by one, at each stage deleting a vertex together with all the edges and faces incident at that vertex. If there were n such edges, the face left after lopping off the pyramid would be an n-sided polygon, which could then be divided into $n - 2$ triangular faces by joining one of the vertices of the polygon to the $n - 3$ vertices not adjacent to it. The resulting polyhedron would have one fewer vertices, three fewer edges, and 2 fewer faces than the original polyhedron, leaving the value of $V - E + F$ unaltered. Continuing in that way, he argued that the number of vertices would eventually be reduced to four, thus leaving a tetrahedron, for which $V - E + F = 2$. That reasoning, however, was flawed. For if applied to the tetrahedron itself (or any pyramid), the procedure does not yield a polyhedron, but a triangulated polygon, for which $V - E + F = 1$; and in other cases the procedure can result in a degenerate polyhedron in which more than two faces share an edge.

Subsequently many other proofs of Euler's formula were given, in attempts not only to rectify Euler's error but to delineate more precisely which polyhedra satisfied the formula and which did not (and indeed, to clarify the very notion of what a polyhedron is). Two of those proofs, by Legendre and Cauchy, are presented in Cromwell (1997) and reproduced below, while Lakatos (1976) provides a detailed critical analysis of Cauchy's proof, a much shorter discussion of Poincaré's abstraction and generalization of Euler's result, and brief mention of Legendre's proof as well as proofs by Joseph Gergonne, Camille Jordan, and Jakob Steiner.[1] One other proof not mentioned in either of those sources is described below.

Legendre restricted consideration to convex solids whose surfaces are composed of polygonal faces. Given such a solid, he mapped its network of edges, via radial projection, onto a corresponding network of arcs of great circles on a unit sphere. Each polygonal face of the original polyhedron then corresponded to a spherical polygon of the same number of sides, and each vertex to a point of intersection of arcs forming such sides, so that the quantities V, E, and F were preserved. He then applied spherical trigonometry to compute the sum of the *areas* of each of the spherical polygons and equated that to the surface area of the sphere. Specifically, the area of a spherical polygon with n sides is given by its spherical excess, the sum of its angles less $(n-2)\pi$. Summing over all the polygons, the total angle sum must be $2\pi V$, since the sum of the angles around each vertex is 2π, and the total number of sides must be $2E$, since each edge is shared by exactly two polygons. Hence

$$4\pi = 2\pi V - 2E\pi + 2\pi F = 2\pi(V - E + F).\ ^{2}$$

Cauchy's proof, in contrast, introduced no metric or other notions foreign to the statement of Euler's theorem. Instead of projecting the network of edges of the polyhedron onto the surface of a sphere, Cauchy imagined removing one face of the polyhedron and deforming the rest of the polyhedron to produce a planar network. Euler's formula would then hold if and only if $V - E + F = 1$ held for that planar configuration. To establish the latter equation, Cauchy triangulated the network by adding diagonals to any non-triangular faces. Every added diagonal increased each of E and F by one, and so left $V - E + F$ invariant. He then systematically described how to remove edges, and if necessary, a vertex, from triangles on the boundary of the network so as to preserve the value of $V - E + F$. If only one edge of such a triangle was on the boundary, he removed just that edge. If two edges lay on the boundary, he removed both of them together with the vertex where they met. In either case, one face was also eliminated. By judiciously[3]

[1]Lakatos (1976), published posthumously, is based on Lakatos's still unpublished and not readily accessible dissertation Lakatos (1961), which, in footnote 2, p. 65 of Lakatos (1976), is said to contain "a detailed heuristic discussion of Euler's, Jordan's and Poincaré's proofs." Lakatos's aim, however, was very different from that of the present inquiry, for he maintained that the tangled history of proofs of Euler's formula and of examples restricting its domain of validity was a paradigm for a general dialectic of "proofs and refutations" involved in the discovery and refinement of all mathematical concepts and theorems. That grandiose claim has been cogently criticized in Feferman (1998); nevertheless, Lakatos (1976) remains a landmark contribution to the philosophy of mathematical practice.

[2]Later Gergonne, and independently, Steiner, derived the same equation by projecting a convex polyhedron onto a plane rather than a sphere and computing the total angle sum of the polygons in the resulting planar network. See Steiner (1826) for details.

[3]As noted in Cromwell (1997), if care is not taken in the order in which the removals are carried out, degenerate networks can result.

repeating the procedure he eventually reduced the network to a single triangular face, for which obviously $V - E + F = 1$.

A proof similar to but distinct from Cauchy's is presented in a particularly perspicuous way in chapter 12 of Rademacher and Toeplitz (1957), a text intended for readers familiar only with high-school algebra and geometry. There the planar network obtained as in Cauchy's proof is regarded as a network of dikes (the edges) protecting fields (the faces) on an island from being flooded by a surrounding sea (representing the face removed at the first step in Cauchy's procedure). To obtain Euler's formula, readers are told to imagine "breaking down one dike after another until all the $F - 1$ fields are under water," taking care "only [to] break dikes that have water on just one side." The end result will then not be a triangle, but a connected network of dikes in which there is a unique path along the dikes from each vertex to any other vertex. (If the network were not connected, some dike surrounded by water on both sides would have to have been destroyed, and if the network contained any loop, the region inside it would not be flooded.) If any particular vertex s is then designated as a starting point, a one-to-one correspondence between the other $V - 1$ vertices and the remaining dikes is given by associating with each such vertex v the last dike traversed along the unique path from s to v. Hence exactly $V - 1$ dikes remain. On the other hand, each time a dike was broken exactly one field was flooded, so $F - 1$ dikes must have been destroyed. The total number E of dikes in the original network must therefore have been $(V - 1) + (F - 1)$.[4]

Ultimately, through the work of Poincaré, Euler's result was transformed into a theorem of algebraic topology concerning simplicial complexes. A version of Poincaré's original proof of that theorem, preceded by definitions of all the prerequisite topological notions, is given on pp. 26–27 of Croom (1978), where the result is stated as

The Euler-Poincaré Theorem: Let K be an oriented [simplicial] complex of dimension n, and for $p = 0, 1, \ldots, n$ let α_p denote the number of p-simplexes of K. Then

$$\sum_{p=0}^{n}(-1)^p\alpha_p = \sum_{p=0}^{n}(-1)^p R_p(K) = \chi(K),$$

where $R_p(K)$ denotes the pth Betti number of K and the integer $\chi(K)$ is the *Euler characteristic* of K.

The Euler-Poincaré formula extends that of Euler in two respects: The alternating sum on the left of the equation generalizes the expression $V - E + F$ to higher dimensions, while the Betti numbers on the right extend the result to polyhedra containing "holes" of various dimensions. The proof is based on dimensional considerations in linear algebra.

---

[4]Note that, in contrast to Cauchy's proof, no triangulation is involved in this argument and no vertices are removed from the network.

13.3   A theorem on rectangular tilings

An article very much in the spirit of the present inquiry is Wagon (1987), which presents and compares fourteen proofs of the following theorem of N.G. de Bruijn.

Theorem: If a rectangle R can be tiled by rectangles each of which has at least one side of integral length, then R must itself have at least one side whose length is an integer.

The proofs considered are based on a variety of different techniques, including (a) double integration (de Bruijn's original method), (b) mathematical induction, (c) polynomials, (d) prime numbers, (e) Eulerian graphs, and (f) step functions. All the proofs are short and are presented in full detail. Three of the proofs are variants of the double integration approach, and two are variants of the inductive method. The others are distinguished from one another not only methodologically, but by whether or not they can be adapted to prove eleven different generalizations of de Bruijn's theorem. Among those generalizations are the following:

(i) Weakening the hypothesis of the theorem by assuming only that for R in standard position (lower left corner at the origin and sides parallel to the coordinate axes), any tiling rectangle that has one corner with integer coordinates has at least one integer side.

(ii) Extending the theorem to n dimensions, to state that if an n-dimensional box B can be filled with n-dimensional boxes each of which has at least one side of integral length, then B must have at least one such side.

(iii) Further extending (ii) to state that if B can be filled with n-dimensional boxes each of which has at least k sides of integral length, then B itself must have at least k sides of integral length.

(iv) Extending the theorem from planar rectangles to rectilinear tilings on a cylinder.

An appendix lists the eleven methodologically distinct proof procedures together with the eleven generalizations of de Bruijn's theorem, and indicates which of the proof procedures can be used to prove each of the generalizations. In particular, of the methods listed above, (c), (e), or (f) can be used to prove (i); any but (b) can be used to prove (ii); any but (b) and (e) to prove (iii); and (b), (e), or (f) to prove (iv). If, following Wagon, "one calls two proofs equivalent provided they work on the same set of generalizations," then, "unless new modifications are found," of the six methods (a)–(f) listed here, generalizations (i)–(iv) rule out the equivalence of any pair except (a) and (d); and in fact, other generalizations show that none of the methodologically distinct approaches are equivalent.

13.4   The geometric-arithmetic mean inequality

For positive numbers a and b, the elementary inequality $\sqrt{ab} \leq \dfrac{a+b}{2}$, with equality if and only if $a = b$, was known in antiquity. More generally, if $a_1, a_2, \ldots, a_n$ and $w_1, w_2, \ldots, w_n$ are two sequences of positive numbers, the *weighted arithmetic mean* $A_n(a, w)$ of $a_1, a_2, \ldots, a_n$ with respect to weights $w_1, w_2, \ldots, w_n$ is defined to be

$$\frac{\sum_{i=1}^{n} a_i w_i}{\sum_{i=1}^{n} w_i},$$

and the *weighted geometric mean* $G_n(a, w)$ of $a_1, a_2, \ldots, a_n$ with respect to $w_1, w_2, \ldots, w_n$ to be

$$\left(\prod_{i=1}^{n} a_i^{w_i} \right)^{1/w},$$

where $w = w_1 + \cdots + w_n$. Then $G_n(a, w) \leq A_n(a, w)$, with equality if and only if $a_1 = \cdots = a_n.$[5]

Even for the elementary inequality there are many proofs, both algebraic and geometric: Six such are given in Nelsen (1993), five more in Nelsen (2000), and nine in the comprehensive monograph Bullen et al. (1988). In the latter source, the justification given for presenting so many proofs of such a simple fact is that "because the definition" of each of those means "is so elementary, there is a certain point in using only the simplest tools to deduce [their] properties" — a concern for purity — while "non-elementary proofs, besides having an interest in themselves, suggest methods of *generalization*, and are often *simpler* than more elementary approaches" (pp. 34–35, italics added). And indeed, three of the proofs of the elementary inequality are later adapted to prove the general result. Before that, however, six proofs are given for the case $n = 2$ and arbitrary weights, followed by proofs that the general result for arbitrary weights follows from that for equal weights and that if, for equal weights, the result holds for some n, then it also holds for all $k \leq n$ (another result due to Cauchy).

In section 2.4 of Bullen et al. (1988), fifty-two proofs of the general theorem are given, "as far as is known, in order of their appearance," beginning with a not fully rigorous proof from 1729 due to Maclaurin[6] and ending with a proof from 1982 by D. Rüthing. In addition to Cauchy, the list includes proofs by such luminaries as Liouville, Hurwitz, Harald Bohr, Richard Bellman, and D.J. Newman. The proofs

---

[5]According to Bullen et al. (1988), this general result "first appeared in print in the nineteenth century," in notes of a course that Cauchy gave at the Ecole Royale.

[6]Maclaurin assumed equal weights and varied the values of $a_1, \ldots, a_n$ while keeping A_n fixed. He assumed without proof that some sequence of a's would maximize the value of G_n.

vary greatly in length. Many are inductive, but others are based on ideas from such disparate branches of mathematics as convex functions, dynamical systems, and even thermodynamics.

13.5   The Law of Quadratic Reciprocity

The Law of Quadratic Reciprocity is one of the cornerstones of algebraic number theory. It concerns quadratic residues, whereby an integer n is said to be a quadratic residue of a prime p if and only if there is an integer x such that $n \equiv x^2 \pmod{p}$. In one version it states:

If p and q are distinct odd primes, then p is a quadratic residue of q if and only if q is a quadratic residue of p, unless $p \equiv q \equiv 3 \pmod 4$, in which case p is a quadratic residue of q if and only if q is not a quadratic residue of p.

A more precise statement was given by Legendre in 1788, for which purpose he introduced the symbol $\left(\frac{p}{q}\right)$, taking the values ± 1, to denote the residue of $p^{(q-1)/2}$ modulo q, where q is an odd prime. In that notation the Law says:

$$\text{For distinct odd primes } p \text{ and } q, \quad \left(\frac{p}{q}\right) = (-1)^{\frac{p-1}{2}\frac{q-1}{2}}\left(\frac{q}{p}\right).$$

Legendre's attempted proof of that Law was incomplete. The first complete proof was found by Gauss in 1796 and published in 1801. In the next seventeen years he published five more proofs of it, and two more were discovered among his unpublished papers.[7] According to Lemmermayer (2000), Gauss was led to devise those alternative proofs in an attempt to find arguments that would generalize so as to yield "similar [laws] for reciprocity of higher powers." Gauss gave his own justification in a remark preceding one of his later proofs of the Law (*Werke* II, 159–160):

> It is a peculiarity of number theory that so many of its most beautiful theorems can be discovered inductively, yet whose proofs lie anything but near at hand, but rather are found only after many fruitless attempts with the help of profound investigations and lucky combinations [of ideas]. This curious phenomenon arises from the often marvelous linkage of different theories in that part of mathematics; and for that very reason it frequently happens that such theorems, a proof of which initially is sought in vain for years, may later be proved in several different ways. As soon as a new theorem is discovered inductively, one must of course regard the finding of any sort of proof to be the first necessity; but after succeeding in that, in number theory one may never consider the investigation to be finished and the tracking down of other proofs to be a superfluous luxury — in part, because one does not usually arrive at first at the most beautiful and simplest proofs, and it is precisely

---

[7]The six published proofs, in German translation, are available online as part of the University of Michigan Historical Math Collection. See http://quod.lib.umich.edu/cgi/t/text/text-idx?c=umhistmath;idno=ABV3158.

the insight into the marvelous linkage of truth in number theory that is what constitutes a principal attraction of that area of study, and that not infrequently leads once again to the discovery of new truths. For these reasons the finding of new proofs for already known truths is here often regarded as at least as important as the discovery of the truth itself.

In the succeeding two centuries the Law of Quadratic Reciprocity has been generalized in many directions, and a multitude of proofs of it have been published. A wide-ranging survey of such generalizations is given in Lemmermayer (2000), together with details of some of the proofs, while an appendix provides a chronological list of 196 published proofs of quadratic reciprocity (not including those in textbooks), together with detailed bibliographic references and brief comments on the methods used. Lemmermayer notes, however (p. 25), that "different entries [therein] do not imply different proofs; some of the proofs are merely reformulations of others. A thorough classification of the published proofs of quadratic reciprocity is something that remains to be done"— a task far beyond the present author's ability to carry out. Readers seeking a challenge may thus wish to consider that project. A place to start is Baumgart (1885). (For a more manageable case study, see item 2. below.)

13.6 Divergence of the series of prime reciprocals

As noted in the summary to Chapter 7 above, it was Euler who first deduced (as a consequence of his product formula) that the series $\sum_{p\ prime}(1/p)$ diverges. Subsequently many other proofs have been given, seven of which are compared in the article Eynden (1980). Among them is a proof by Erdős "notable for its lack of series manipulations," which, though indirect, does not invoke the divergence of the harmonic series, and two proofs by Eynden himself, the first of which is a simple direct proof. Of the other four proofs discussed, three (by Richard Bellman, Erich Dux, and James Clarkson) involve rearrangement of the terms of an infinite series and the fourth (by Leo Moser) uses the theorem that a convergent series is also Cesàro summable to the same limit. Still another proof, by Frank Gilfeather and Gary Meisters, is noted but not discussed. Eynden's article includes source references for all the proofs mentioned therein.

13.7 The brachistochrone problem

In June 1696 Johann Bernoulli famously challenged other mathematicians to determine, given two points A and B in a vertical plane, the curve in that plane down which a frictionless bead would travel from A to B under the influence of gravity in the least time (a curve he called the brachistochrone), and promised to publish his own solution of the problem by the end of the year if no one else was able to find the answer.

That same month, however, within a week of becoming aware of the problem, Leibniz wrote to Bernoulli with a sketch of his derivation of an equation for that curve. Three months later Leibniz then wrote to Bernoulli's older brother, Jakob, urging him to work on the problem, and by early October Jakob, too, had found the solution. Then in January of 1697 Isaac Newton published an anonymous, very terse note in which he stated without proof that the brachistochrone is the arc of the cycloid passing through A and B, and the following May Johann Bernoulli published both his solution and that of his brother.[8]

The proofs of the two Bernoullis are excerpted in English translation in Struik (1969). That by Johann is an ingenious physical argument: it proceeds by translating the question into a corresponding problem in optics (finding the path followed by a light ray in a laminar medium of continuously varying index of refraction), using Fermat's principle that light always travels along the path of least time, which implies Snell's law of refraction. Jakob's derivation, on the other hand, was purely analytic in character. Though less elegant than Johann's, Jakob's approach was more widely applicable to other problems in the calculus of variations.

Chapter 1 of Goldstine (1980) includes a detailed comparison of the derivations of Leibniz, Newton, and the two Bernoullis.[9] They serve as exemplars of alternative proofs arising from differences in individual patterns of thought, since each was obtained in ignorance of the others.

Some further case studies worth pursuing

Theorems having only one known proof are very much the exception rather than the rule. There is thus wide scope for comparative studies of alternative proofs. The following brief list suggests a few among many possibilities for further investigation.

1. **Quadrature of the parabola**. In Arana and Mancosu (2012), pp. 297–299, it is mentioned that Evangelista Torricelli, in his 1644 treatise *De quadratura parabolae* (part of the *Opera Geometrica* in Torricelli 1919–1944) gave no less than twenty different derivations of the area of a parabolic segment bounded by a parabolic arc and a chord of the parabola. The result (that the area is $4/3$ that of the triangle whose base is the chord and whose opposite vertex is the point on the parabola where the tangent is parallel to the chord) was first obtained by Archimedes in his own *Quadrature of the Parabola*, using geometric methods and a double *reductio* argument. Torricelli's proofs include some of that type, and others based on Cavalieri's method of indivisibles.

---

[8]These historical details are taken from the account in chapter 1 of Goldstine (1980).

[9]Further on in that chapter Goldstine also discusses a second solution of the brachistochrone problem by Johann Bernoulli. Published in 1718, it was based on an idea that foreshadowed a powerful technique in the calculus of variations later developed by Constantin Carathéodory.

2. **Fermat's theorem on the sum of two squares**. One of Fermat's famous results is that every prime of the form $4m + 1$ (of which there are infinitely many) is a sum of two squares. Section 3.1 of Weintraub (2008), entitled "Fermat's Theorem," gives three different proofs of that theorem; Aigner and Ziegler (2000) gives two different proofs of the result, the first due to Axel Thue and the second to Roger Heath-Brown; three different proofs are also given in section 2.3 of Avigad (2006), an article concerned with issues very similar to those considered here; and Lemmermayer (2000), p. 12, refers to several more. The methods employed vary widely, and include the theory of Euclidean rings, continued fractions, and the geometry of numbers.

3. **Morley's Theorem**. The theorem states that in any triangle the three points of intersection of adjacent angle trisectors are the vertices of an equilateral triangle. Coxeter (1961), pp. 24–25, presents one proof and gives references to several others, both geometric and trigonometric; and very recently John Conway has published what he calls "the indisputably simplest proof" (Conway 2014).

4. **The Erdős-Mordell inequality**. Let P be any point inside a triangle ABC, let r_a, r_b, and r_c be the perpendicular distances from P to each of the sides a, b, c of the triangle, and let R_A, R_B, and R_C be the distances from P to each of the vertices A, B, C. Then $R_A + R_B + R_C \geq 2(r_a + r_b + r_c)$. The paper Komornik (1997) gives a very elementary proof and cites a number of other earlier proofs.

5. **The fundamental theorem of calculus**. Proofs of the fundamental theorem of calculus depend upon what notion of area or integral is considered and what the underlying space is. The standard theorem in the context of Riemann integration on the real line can, for example, be extended to the Lebesgue differentiation theorem in one direction and to Stokes Theorem in another, and the latter can be expressed in down-to-earth terms or in the abstract setting of differential forms. There are many possibilities for investigation and many possible sources.

6. **Sylow's theorems in group theory**. As noted in the Wikipedia entry on that subject, "The Sylow theorems have been proved in a number of ways, and the history of the proofs themselves are the subject of many papers." In particular, one may consult Casadio and Zappa (1990), Gow (1994), Waterhouse (1980), and Wielandt (1959).

7. **The fundamental theorem of Galois theory**. There are at least two distinct proofs of the fundamental theorem of Galois theory, one due to Emil Artin and the other invoking the primitive element theorem. A comparison of their merits (and those of other possible approaches) might be worthwhile.

I look forward to whatever further studies and discussion may be stimulated by the examples and points of view I have presented in this book.

References

Aigner, M., Ziegler, G.M.: Proofs from the Book, 2nd ed. Springer, Berlin (2000)

Arana, A., Mancosu, P.: On the relationship between plane and solid geometry. Rev. Symb. Logic **5**(2), 249–353 (2012)

Avigad, J.: Mathematical method and proof. Synthese **193**(1), 105–159 (2006)

Baumgart, O.: Ueber das quadratische Reciprocitätsgesetz. Eine vergleichende Darstellung der Beweise. Dissertation, U. Göttingen. Zeitschrift Math. Phys. **30** (1885)

Bullen, P.S., Mitrinović, D.S., Vasić, Means and Their Inequalities. D. Reidel, Dordrecht et al. (1988)

Casadio,G.,Zappa,G.: History of the Sylow theorem and its proofs. Boll. Storia Sci. Mat. **10**, 29–75 (1990)

Conway, J.: On Morley's trisector theorem. Math. Intelligencer **36**(3), 3 (2014)

Coxeter, H.S.M.: Introduction to Geometry. Wiley, New York (1961)

Cromwell, P.: Polyhedra. Cambridge U.P., Cambridge (1997)

Croom, F.: Basic Concepts of Algebraic Topology. Springer-Verlag, New York, Heidelberg and Berlin (1978)

Dunham, W.: An ancient/modern proof of Heron's formula. Math. Teacher **78**, 258–259 (1985)

Dunham, W.: Journey through Genius: The Great Theorems of Mathematics. Wiley, New York (1990)

Dunham, W.: Euler, the Master of Us All. Math. Assn. Amer., Washington, D.C. (1999)

Dunham, W.: Newton's proof of Heron's formula. Math Horizons **19**, 5–8 (2011)

Eynden, C.V.: Proofs that the series of prime reciprocals diverges. Amer. Math. Monthly **5**, 394–397 (1980)

Feferman, S.: The logic of mathematical discovery versus the logical structure of mathematics. In S. Feferman, In the Light of Logic, pp. 77–93. Oxford U.P., New York and Oxford (1998)

Goldstine, H.: A History of the Calculus of Variations from the 17th through the 19th Century. Springer, New York et al. (1980)

Gow, R.: Sylow's proof of Sylow's theorem. Irish Math. Soc. Bull. **33**, 55–63 (1994)

Komornik, V.: A short proof of the Erdős-Mordell theorem. Amer. Math. Monthly **104**, 57–60 (1997)

Lakatos, I.: Essays in the logic of mathematical discovery. Unpublished Ph.D. dissertation, Cambridge U. (1961)

Lakatos, I.: Proofs and Refutations: The Logic of Mathematical Discovery (ed. J. Worrall and E. Zahar). Cambridge U.P., Cambridge (1976)

Lemmermayer, F.: Reciprocity Laws from Euler to Eisenstein. Springer, Berlin et al. (2000)

Nelsen, R.: Proofs without Words. Math. Assn. Amer., Washington, D.C. (1993)

Nelsen, R.: Proofs without Words II. Math. Assn. Amer., Washington, D.C. (2000)

Oliver, B.: Heron's remarkable triangular area formula. Math. Teacher **86**, 161–163 (1993)

Rademacher, H., Toeplitz, O.: The Enjoyment of Mathematics. Princeton U.P., Princeton (1957)

Steiner, J.: Leichter Beweis eines stereometrischen Satzes von Euler. J. reine und angewandte Math. **1**, 364–367 (1826)

Struik, D. (ed.): A Source Book in Mathematics, 1200–1800. Harvard U.P., Cambridge (1969)

Torricelli, E.: Opere di Evangelista Torricelli (ed. G. Loria and G. Vassura). Stabilimento Tipografico Montanari, Faenza (1919–1944)

Wagon, S.: Fourteen proofs of a result about tiling a rectangle. Amer. Math. Monthly **94**, 601–617 (1987)

Waterhouse, W.: The early proofs of Sylow's theorem. Arch. Hist. Exact Sci. **21**, 279–290 (1980)

Weintraub, S.H.: Factorization: Unique and Otherwise. AK Peters, New York (2008)

Wielandt, H.: Ein Beweis für die Existenz der Sylowgruppen. Arch. Math. **10**, 401–402 (1959)

Erratum

Why Prove it Again?

Alternative Proofs in Mathematical Practice

John W. Dawson, Jr.

© Springer International Publishing Switzerland 2015
J.W. Dawson, Jr., *Why Prove it Again?*,
DOI 10.1007/10.1007/978-3-319-17368-9

DOI 10.1007/978-3-319-17368-9_14

The publisher regrets the errors in the print and online versions of this book. Corrections to chapters 7, 9 and 11 can be found on the next page.

© Springer International Publishing Switzerland 2015
J.W. Dawson, Jr., *Why Prove it Again?*, DOI 10.1007/978-3-319-17368-9_14

p. 56, line 6, insert comma after 1955.

p. 99, lines −8 and −9, insert x in the numerators of the fractions in the equations for the second of each pair of lines.

p. 101, line 19, $A1B_1C_1$ should be $A_1B_1C_1$.

p. 156, lines 1, 2, 6, and 9, add the subscript k to each unsubscripted F.

p. 157, lines 8 through 10, the subscript on Φ should be p^k, not p.

p. 158, line 2, "$= 1$" should be "$= p$".

p. 160, line 1, the leading term of the polynomial should be deleted and the penultimate term should be $x^{(p^{k-1})}$, in agreement with the definition of $\Phi_{(p^k)}(x)$ on page 150.

p. 161, line 7, $x^n - 1$ should be $x^q - 1$.

p. 161, lines 13 and 15, m^2 should be q^2 in all three occurrences.

p. 162, line 9, p^k should be p^j in all three occurrences.

p. 166, line 15, $\Phi(m)$ should be $\varphi(m)$, where φ denotes Euler's totient function (cf. p. 150).

The online version of the original book can be found at
http://dx.doi.org/10.1007/978-3-319-17368-9

Index

© Springer International Publishing Switzerland 2015
J.W. Dawson, Jr., *Why Prove it Again?*, DOI 10.1007/978-3-319-17368-9